# Lecture Notes in Mathematics

Edited by A. Dold and B. Eckmann



720

Ed Dubinsky

# The Structure of Nuclear Fréchet Spaces

Springer-Verlag
Berlin Heidelberg New York 1979

**Author**

Ed Dubinsky
Mathematics Department
Clarkson College
of Technology
Potsdam, N.Y. 13676
USA

AMS Subject Classifications (1970): 41A65, 46-02, 46A05, 46A15, 46A35, 46A45, 47B05, 47B10

ISBN 3-540-09504-7 Springer-Verlag Berlin Heidelberg New York
ISBN 0-387-09504-7 Springer-Verlag New York Heidelberg Berlin



2141/3140-543210

To Andrzeja

"We have the greatest freedom when we are bound together with those we love"

TABLE OF CONTENTS

# INTRODUCTION

These notes are based on two series of lectures. The first was given at Helsinki University of Technology, Otaniemi, Finland during April and May, 1978. The second was given at Universität Kaiserslautern in October, 1978 and repeated at Gesamthochschule Wuppertal the following month.

In recent years there has been a fair amount of interest in the theory of nuclear Fréchet spaces. Most of the research which has been done can be divided into three classifications: approximation properties including bases and finite dimensional approximations of the identity; the question of when a nuclear Fréchet space F is isomorphic to a subspace, quotient space or complemented subspace of a given nuclear Fréchet space E; and the analysis of various function spaces which are nuclear Fréchet spaces, such as spaces of holomorphic functions (on both finite and infinite dimensional domains) and spaces of $C^{\infty}$-functions.

These notes are concerned with only the first two classifications. Essentially everything that is known about approximation properties is presented. Regarding the second category we confine ourselves to the case in which F has a basis. Thus we omit (except for references) the beautiful and powerful basis free theory which is now being developed. In our situation we obtain various characterizations in terms of certain surprisingly simple inequalities. These formulas and the techniques involved in obtaining them have already proved useful in other contexts (such as approximation properties and spaces of holomorphic functions) and should find further applications.

Some specific topics have been omitted. These include questions related to the quasi-equivalence problem and the existence of non-compact operators. The only reasons for this are the usual difficulties of time and space.

One major reason for producing these notes is the hope that they will stimulate further research. For this reason I have tried to indicate specific problems and areas where I think further work could be fruitful. Already there are indications that this is happening. For example, some results of H. Ahonen, which were obtained too late for inclusion here, tell us a bit more about the material in Chapter II. If these notes do help some mathematicians to obtain more significant information about these questions, then I will feel that they are successful. For me it is a fascinating study and I extend an open invitation to join in.

The material is divided into six chapters. The last section in each chapter is entitled, Notes and Remarks. Here are included bibliographical references, miscellaneous comments and an indication of which results are not published elsewhere. These are results which, although important and of some interest, are perhaps too technical for publication in a journal. The Lecture Notes series seems to be an excellent medium for such material.

In Chapter I we develop notation and many preliminary results. Usually our characterizations of subspaces and quotient spaces involve two conditions. One of them is a "strength of nuclearity" condition expressed in terms of the asymptotic behavior of the Kolmogorov diameters. All results involving the necessity of this condition are immediate consequences of theorem (6.2.4). Another important feature of this chapter is propositions (5.3) and (5.4) which greatly simplify the rather complicated calculations involved in the study of $L_f(\alpha,r)$ spaces.

In Chapter II we study the other condition involved in the characterizations. This is an inequality involving the matrix which represents the basis in a nuclear Fréchet space. So far there are eight such inequalities which have been discovered—the $(d_i)$ conditions,

$i = 1,\ldots,6$ studied in this chapter and two others introduced later in connection with the $L_f(\alpha,r)$ spaces. Here we study the various properties of the $(d_i)$ conditions and some consequences. Propositions (2.2.3) and (2.3.3) give the main idea used in the construction of subspaces and quotient spaces respectively. This idea appears again in the study of approximation properties. Finally, we give some constructions which are not characterizations but do provide important structural information about nuclear Fréchet spaces.

Chapter III is the heart of our work on subspaces. After deriving the fundamental inequality which gives all results involving the necessity of the other condition in the characterization, we systematically describe the construction of subspaces with bases in stable power series spaces (of infinite and finite type) unstable power series spaces, weakly stable power series spaces, mixed power series spaces (that is, Cartesian products of spaces of different type) and stable $L_f(\alpha,r)$-spaces. In the first and the last case we obtain complete characterizations. In the unstable and weakly stable case there is a lot of information and many gaps. We really don't know anything at this time about mixed spaces, although some recent investigations of K. Nyberg are promising.

In Chapter IV we attempt to dualize our subspace constructions to obtain quotient spaces. The method is only partially successful and so we only consider the case of stable power series spaces. Some consequences of these results are derived.

Chapter V serves as a transition between our investigation of subspaces and quotient spaces and the study of approximation properties. In the first part we analyze the nature of a complemented subspace under the assumption that it has a basis. In the second part we show that in certain cases a complemented subspace of a space with a basis automatically has a basis. Actually, it is not necessary to begin with a complemented subspace. We obtain our result only assuming that F is isomorphic to a subspace of E and to a quotient space of E. Thus it would be interesting to know exactly when this implies that F is isomorphic to a complemented subspace of E.

In Chapter VI we present essentially all that is known about approximation properties in nuclear Fréchet spaces. We begin with a definition of each of the usual properties. Then we completely answer the question of whether every nuclear Fréchet space has each property (existence problem) and develop a considerable amount of information about whether these properties are preserved under subspaces and quotient spaces (permanence problem) and about which implications between pairs of properties are valid (comparison problem).

These notes were written during calendar year 1978 while I was on sabbatical leave from Clarkson College of Technology. During that time I was partially supported by a grant from the National Science Foundation.

I spent my sabbatical traveling to many different places and so I am grateful to several institutions for their hospitality. In particular, I would like to thank the Banach Institute, Polish Academy of Sciences where much of the writing was done and also the three institutions where the lectures were given.

The original suggestion that I write these notes came from H. Jarchow and I would like to thank him for his encouragement in this work.

In Helsinki, my hosts were K. Vala and H. Apiola and the gratitude that I and my family feel for their kind hospitality is not easy to express. The functional analysis group in Helsinki was created by Vala and I think his contribution to mathematics as a teacher and an inspiration to young mathematicians deserves to be recognized.

It was necessary to train the students in Helsinki in the preliminary material for several months before my arrival. This was done by Apiola and because of his efforts my task was greatly eased.

My hosts in Kaiserslautern and Wuppertal were E. Schock and D. Vogt, respectively. I would like to thank them both for their hospitality.

I would like to thank my listeners, and in particular I would like to acknowledge the assistance of H. Ahonen, L. Holmstrom, T. Ketonen and K. Nyberg who looked (with much success) for errors and made many suggestions for improvements.

My attitudes toward this work and to functional analysis in general have been greatly influenced by my interactions with C. Bessaga, P. Djakov, B. Mitiagin and A. Pełczyński. I would like to express my gratitude to all of them.

Finally, I would like to thank my wife Andrzeja and our three daughters, Malgosia, Hanna and Ewa. They wandered around with me during most of this sabbatical year—and that was not easy. But their presence and understanding was a major contribution to this effort. I hope the result is worthy of them.

# CHAPTER I

## PRELIMINARIES

### 1. Prerequisites

(1.1) We assume familiarity with the general theory of locally convex spaces as found, for example in [42] or [57] and the elementary theory of nuclear spaces as found in [53]. Actually, very little of this is needed. Only a few definitions and basic results, most of which are discussed in this chapter will be used. Any other results which are used will either be described in detail or be given a specific reference in the literature.

(1.2) All references attributing various results to specific authors will be found in the notes and remarks section at the end of each chapter. Sometimes references to proofs and definitions which are omitted (especially in this chapter) will also be included with the notes and remarks. Most proofs in this first chapter are omitted.

(1.3) The expression II (3.7) will refer to subsection 7 of section 3 in Chapter II. When this reference appears within Chapter II it will be shortened to (3.7). The numbering system is otherwise self explanatory.

### 2. Terminology

(2.1) We denote by $\mathbb{N}$ the sequence of positive integers. By the phrase, a permutation $\pi$ of $\mathbb{N}$ we mean a bijection $\pi: \mathbb{N} \to \mathbb{N}$. The word neighborhood will be abbreviated, nbd. Sometimes we will consider the ratio of two numbers when it is possible that the denominator is 0. In every case it then follows that the numerator is also 0 and the value of the ratio is taken to be 0.

(2.2) By a subspace of a topological vector space E we mean a closed, infinite dimensional vector subspace, unless otherwise specified. The topology is the usual induced topology. By a quotient space of E we mean the image of a continuous linear surjection with the usual quotient space topology. By a complemented subspace of E we mean a subspace F such that $\exists$ a not necessarily infinite dimensional subspace G with $F \cap G = \{0\}$ and $E = F + G$.

(2.3) An isomorphism is a linear bijection which is continuous and open.

(2.4) By $\omega$ we denote the space of all sequences of (real or complex) scalars with the product topology.

(2.5) If E is a vector space and $||\cdot||$ is a norm on E then $(E,||\cdot||)^{\wedge}$ will denote the Banach space obtained by completing E with respect to the norm $||\cdot||$.

3. Köthe Spaces

(3.1) A basis in a topological vector space E is a sequence $(x_n)$ in E such that for each $x \in E$ there is a unique sequence of scalars $(t_n)$ such that $x = \sum_n t_n x_n$. A basic sequence is a sequence which is a basis in the subspace it generates. If $(x_n)$ is a basis in E, $(p_n)$ a strictly increasing sequence of integers with $p_0 = 0$ and $y_n \neq 0$ has the form

$$y_n = \sum_{i=p_{n-1}+1}^{p_n} t_i x_i$$

where $(t_i)$ is a sequence of scalars, then $(y_n)$ is called a block basic sequence with respect to the basis $(x_n)$. It is easy to see that $(y_n)$ is a basic sequence.

(3.2) If E is a nuclear Fréchet space with a basis $(x_n)$ and $(||\cdot||_k)$ is an increasing sequence of seminorms which defines the topology of E then we have an infinite matrix $a = (a_n^k)$ where $a_n^k = ||x_n||_k$, $k, n \in \mathbb{N}$. We define the Köthe space $K(a)$ corresponding to E, $(x_n)$ to be

$$K(a) = \{\xi=(\xi_n): |\xi|_k = \sum_n |\xi_n| a_n^k < \infty, k \in \mathbb{N}\} \quad (1)$$

with topology defined by the sequence of seminorms, $(|\cdot|_k)$. Clearly K(a) is independent of the choice of $(||\cdot||_k)$, but a is not.

It is also clear that if the value of $a_n^k$ is changed in such a way that for each k there is $n_k$ such that $a_n^k$ is unchanged for $n \geq n_k$ then K(a) is also unchanged both as a set and in its topology. Such a procedure, which is used extremely often, is called an <u>adjustment</u> of a.

K(a) is a nuclear Fréchet space. Moreover, if $e_n$ is the sequence which is 0 at each coordinate except the nth where it is 1, then $(e_n)$ is a basis for K(a). We call it the <u>coordinate basis.</u> The map which sends $e_n$ to $x_n$ has a unique extension to an isomorphism of K(a) onto E.

The matrix of a is called a <u>representation of the basis</u> $(x_n)$. It satisfies:

$$0 \leq a_n^k \leq a_n^{k+1}, \quad \sup_k a_n^k > 0, \forall k \exists j \ni \sum_n \frac{a_n^k}{a_n^j} < \infty \quad (2)$$

(3.3) Conversely if a is an infinite matrix satisfying (2) and K(a) is defined by (1) then K(a) is a nuclear Fréchet space with basis. It is in this sense that we say that "nuclear Köthe space" means the same thing as nuclear Fréchet space with basis.

(3.4) Given K(a) as defined in (1) with a satisfying (2) it follows that the topology of K(a) is also defined by the sequence of seminorms, $(||\cdot||_k)$ where

$$||\xi||_k = \sup_n |\xi_n| a_n^k, \quad \xi \in K(a)$$

We call $(\|\cdot\|_k)$ the <u>sequence of sup norms</u> relative to the matrix a.

(3.5) If K(a) is a Köthe space and $\mathbb{N}_1 \subset \mathbb{N}$ then the <u>stepspace</u> $(K(a))_{\mathbb{N}_1}$ is the Köthe space K(b) where $b^k = a^k|_{\mathbb{N}_1}$. That is, it is obtained by taking each sequence in K(a) and restricting it to $\mathbb{N}_1$.

(3.6) If K(a) and K(b) are two Köthe spaces and $t = (t_n)$ is a sequence of positive numbers such that

$$K(b) = \{(t_n \xi_n): (\xi_n) \in K(a)\}$$

then we say that K(b) is a <u>diagonal transform</u> of K(a)

## 4. Power Series Spaces

(4.1) Let $\alpha = (\alpha_n)$ be an increasing sequence of positive numbers. We say that $\alpha$ is a <u>nuclear exponent sequence of infinite type</u> if

$$\sup_n \frac{\log n}{\alpha_n} < \infty \tag{3}$$

In this case we define the <u>infinite type power series space</u> $\Lambda_\infty(\alpha)$ to be K(a) where $a_n^k = e^{k\alpha_n}$. It is clear in our context that (3) is equivalent to (2).

(4.2) Let $\alpha = (\alpha_n)$ be an increasing sequence of positive numbers. We say that $\alpha$ is a <u>nuclear exponent sequence of finite type</u> if

$$\lim_n \frac{\log n}{\alpha_n} = 0 \tag{4}$$

In this case we define the <u>finite type power series space</u> $\Lambda_1(\alpha)$ to be K(a) where $a_n^k = e^{-\frac{1}{k}\alpha_n}$. It is clear in our context that (4) is equivalent to (2).

(4.3). When there is no possibility of confusion we will write $\Lambda_\infty(\alpha)$, $\Lambda_1(\beta)$ assuming implicitly that $\alpha$ is a nuclear exponent sequence of infinite type and $\beta$ is a nuclear exponent sequence of finite type.

(4.4) If $\alpha_n = \log n$ we will denote the power series space $\Lambda_\infty(\alpha)$ by (s), the space of rapidly decreasing sequences.

(4.5) Lemma

Let $\alpha, \beta$ be nuclear exponent sequences of infinite type and assume that

$$\sup_n \frac{\alpha_{2n}}{\alpha_n} = A < \infty \ , \ \sup_n \frac{\alpha_n}{\beta_n} < \infty$$

Then there is a strictly increasing sequence of indices $(j_n)$ such that

$$0 < \inf_n \frac{\alpha_{j_n}}{\beta_n} \leq \sup_n \frac{\alpha_{j_n}}{\beta_n} < \infty$$

Proof

We may assume that $\alpha_n \leq \beta_n$, $n \in \mathbb{N}$. By (3), $\lim_n \alpha_n = \infty$ so we can find integers $(k_n)$ with $\alpha_{k_n - 1} \leq \beta_n < \alpha_{k_n}$ and since $\beta_n \leq \beta_{n+1}$ it follows that $n \leq k_n \leq k_{n+1}$. Let $j_n = k_n + n$ so $j_n < j_{n+1}$ and

$$1 < \frac{\alpha_{k_n}}{\beta_n} \leq \frac{\alpha_{j_n}}{\beta_n} \leq \frac{\alpha_{2k_n}}{\beta_n} \leq A \frac{\alpha_{k_n}}{\beta_n} = A \frac{\alpha_{k_n}}{\alpha_{k_n - 1}} \frac{\alpha_{k_n - 1}}{\beta_n} \leq A^2$$

which gives the desired result. ∎

## 5. The Spaces $L_f(\alpha,r)$

(5.1) Let f be a real valued function of a real variable. We assume that f is continuous strictly increasing, odd and logarithmically convex on the non-negative reals—that is, $\log f(e^x)$ is a convex function of x when $x \geq 0$. We also assume that f is rapidly increasing—that is,

$$\lim_{x\to\infty} \frac{f(\rho x)}{f(x)} = \infty \text{ for each } \rho > 1 .$$

Let $\alpha = (\alpha_n)$ be an increasing sequence of positive numbers and let $r\varepsilon(-\infty,\infty]$. We assume that there exists $\rho < r$ such that

$$\sum_{n=1}^{\infty} e^{-f(\rho\alpha_n)} < \infty .$$

With these assumptions we define $L_f(\alpha,r)$ to be the Köthe space K(a) where $a_n^k = e^{f(\rho_k\alpha_n)}$ where $(\rho_k)$ is a strictly increasing sequence which converges to r.

(5.2) Proposition

In the context of (5.1) we have,

i) $L_f(\alpha,r)$ is a nuclear space

ii) Each $L_f(\alpha,r)$ space is isomorphic to one in which r has one of the four values: -1, 0, 1, $\infty$.

iii) The ratio $\dfrac{a_n^k}{a_n^{k+1}}$ is decreasing with n for each k.

(5.3) Proposition

In the context of (5.1) we have

i) Given $0 < \delta < 1 < B$ there exists $x_0 \geq 0$ such that

$$\delta f(y) \leq f(y) - f(x) \text{ whenever } x_0 \leq x \leq Bx \leq y$$

ii) Given positive numbers $M,\varepsilon$ there exists $x_0 \geq 0$ such that

$$f^{-1}(x) \geq (1-\varepsilon)f^{-1}(Mx), \quad x \geq x_0$$

iii) If $B > 1$ then

$$\lim_{x\to\infty} \frac{f^{-1}(Bx)}{f^{-1}(x)}$$

Proof

i) Given $\delta, B$ we apply the fact that f is rapidly increasing to obtain $x_0 \geq 0$ such that

$$\frac{f(Bx)}{f(x)} \geq \frac{1}{1-\delta} \text{ for } x \geq x_0$$

Then, since f is increasing, if $x_0 \leq x \leq Bx \leq y$ we have

$$(1-\delta)f(y) \geq (1-\delta)f(Bx) \geq f(x)$$

which is the desired relation.

ii) First we observe that since f is continuous, strictly increasing and not bounded above or below then $f^{-1}$ is defined, continuous, strictly increasing and not bounded.

If $\varepsilon \geq 1$ the result is clear so we may assume that $0 < \varepsilon < 1$. Now apply the fact that f is rapidly increasing to find $y_0$ such that

$$f(\tfrac{1}{1-\varepsilon}y) \geq Mf(y) \text{ for } y \geq y_0$$

Applying $f^{-1}$ to both sides and writing $x = f(y)$ we obtain the desired inequality valid for $x \geq x_0 = f(y_0)$.

iii) Since $B > 1$ and $f^{-1}$ is increasing we have

$$\frac{f^{-1}(Bx)}{f^{-1}(x)} > 1 \text{ for } x \geq 0$$

On the other hand, given $\delta > 0$ choose $\varepsilon > 0 \ni$ $0 < \frac{1}{1-\varepsilon} \leq 1 + \delta$. Then applying ii) we have $x_0 \ni$

$$\frac{f^{-1}(Bx)}{f^{-1}(x)} \leq \frac{1}{1-\varepsilon} \leq 1 + \delta \text{ for } x \geq x_0$$

∎

(5.4) Proposition

Let g be any function satisfying the conditions of (5.1) and let $\alpha, r$ be as in (5.1). Then there exists a function f satisfying the conditions of (5.1) such that $L_f(\alpha,r) = L_g(\alpha,r)$ and there exists $R > 1$ such that

$$f^{-1}(My) \leq MRf^{-1}(y) \text{ for all } M \geq 1,\ y \geq 0 \quad . \qquad (1)$$

Proof

We will construct f so that $f(x) = g(x)$ for $|x|$ sufficiently large. This will guarantee that $L_f(\alpha,r) = L_g(\alpha,r)$.

Since g is rapidly increasing, we have $x_0 \ni g(2x) \geq 2g(x)$ for all $x \geq x_0$ and so for any positive integer n and $x \geq x_0$ we have

$$g(2^n x) \geq 2^n g(x) \quad .$$

Hence if $M \geq 1 \ \exists\ n \ni 2^{n-1} \leq M < 2^n$ and we have for $x \geq x_0$

$$g(2Mx) \geq 2g(Mx) \geq 2g(2^{n-1}x) \geq 2^n g(x) > Mg(x) \quad .$$

Therefore it follows by taking inverses that

$$g^{-1}(My) \leq 2Mg^{-1}(y) \text{ for all } M \geq 1,\ y \geq y_0 = g(x_0)\ . \tag{2}$$

It also follows from this argument that there exists $\delta > 0$ ∋

$$g(x) \geq \delta x \text{ for } x \geq x_0$$

and since g is rapidly increasing we can find $x_1 \geq \log x_0$ ∋

$$g(x) \geq xg(1) \text{ for } x \geq e^{x_1}\ . \tag{3}$$

Now we define f by

$$f(x) = \begin{cases} g(1)x^{\frac{1}{x_1}\log\frac{g(e^{x_1})}{g(1)}} & \text{if } 0 \leq x \leq e^{x_1} \\ g(x) & \text{if } x \geq e^{x_1} \\ -f(-x) & \text{if } x \leq 0 \end{cases}$$

The function $\log f(e^x)$ is logarithmically convex because its graph is obtained by taking the graph of $\log g(e^x)$, $x \geq x_1$ and connecting the points $(0, \log g(1))$ and $(x_1, \log g(e^{x_1}))$ in the plane by a straight line. Thus f satisfies the conditions of (5.1) and since $f(x) = g(x)$ for $|x| \geq e^{x_1}$ it follows that $L_f(\alpha,r) = L_g(\alpha,r)$ no matter what are $\alpha, r$.

Finally we must verify (1). Set $y_1 = g(e^{x_1})$, $a = \frac{1}{g(1)}$, $b = \frac{y_1}{g(1)}$. Then we have

$$f^{-1}(y) = \begin{cases} (ay)^{\frac{x_1}{\log b}} & \text{if } 0 \le y \le y_1 \\ g^{-1}(y) & \text{if } y \ge y_1 \end{cases}$$

Suppose that $M \ge 1$. If $y \ge y_1$ then $My \ge y_1$ and the desired inequality is just (2) provided $R \ge 2$.

If $My \le y_1$ then $y \le y_1$ and it follows from (3) that $g(e^{x_1}) \ge g(1)e^{x_1}$, that is, $x_1 \le \log b$ and so

$$f^{-1}(My) = (aMy)^{\frac{x_1}{\log b}} \le M(ay)^{\frac{x_1}{\log b}} = Mf^{-1}(y)$$

and (1) holds provided $R \ge 1$.

Finally, suppose $y < y_1 < My$. Then from what we have shown,

$$f^{-1}(My) = f^{-1}(\frac{My}{y_1}y_1) \le 2\frac{My}{y_1}f^{-1}(y_1) = 2\frac{My}{y_1}f^{-1}(\frac{y_1}{y}y) \le 2Mf^{-1}(y) \quad .$$

∎

(5.5) If f is a function which satisfies all of the conditions of (5.1) for some $\alpha, r$ and relation (1) of (5.4) holds for some $R > 1$, then we shall say that f is a <u>Dragilev function</u> with respect to $\alpha, r$. Whenever we mention $L_f(\alpha, r)$ it is implicitly assumed that f is a Dragilev function with respect to $\alpha, r$.

## 6. Kolmogorov Diameters

(6.1) Preliminaries

(6.1.1) Let U,V be absolutely convex sets in a vector space E such that U absorbs V. We define the nth <u>Kolmogorov diameter</u> of V with respect to U, $n = 0,1,2,\ldots$, by

$$d_n(V,U) = d_n(V,U;E) = \inf_{L\in F_n} \inf\{\delta>0:\ V\subset\delta U+L\}$$

$$= \inf_{L\in F_n} \sup_{x\in V} \inf_{y\in L} ||x-y||_U ,$$

where $F_n = F_n(E)$ is the set of all subspaces of E of dimension $\leq n$ and $||\cdot||_U$ is the gage of U, that is,

$$||x||_U = \inf\{\lambda>0:\ x\in\lambda U\} .$$

The equality of the two expressions for $d_n(V,U)$ is elementary.

(6.1.2) We will use the following elementary properties of diameters

i) If $V_1\subset V$ and $U\subset U_1$ then $d_n(V_1,U_1) \leq d_n(V,U)$

ii) If s,t are positive numbers then $d_n(sV,tU) = \frac{s}{t}d_n(V,U)$

iii) If $T:E \to F$ and $S:F \to G$ are linear maps on vector spaces and U,V,W are absolutely convex sets in E,F,G respectively, such that V absorbs T(U) and W absorbs S(V) then

$$d_{n+m}(ST(U),W) \leq d_n(T(U),V)d_m(S(V),W)$$

iv) If F is a vector subspace of E and $V\cup U\subset F$ then

$$d_n(V,U;F) = d_n(V,U;E)$$

(6.2) Permanence with Respect to Subspaces and Quotient Spaces

(6.2.1) In the literature some of the characterizations of subspaces are expressed in terms of various stronger forms of nuclearity. It may be simpler to write the conditions directly in terms of the asymptotic behavior of the diameters. We are able to do this with the help of one theorem for which we need two lemmas.

(6.2.2) Lemma

Let U,V be absolutely convex nbds of 0 in a locally convex space E such that U is closed and absorbs V and the gage of U is a norm which is defined by an inner product. Let F be a vector subspace (not necessarily closed) of E.

Then

$$d_n(V\cap F, U\cap F) \leq d_n(V,U) \quad \forall n \quad .$$

Proof

Let $\hat{E}_U = (E,||\cdot||_U)^\wedge$ and if S is a subset of $\hat{E}_U$, denote by $\hat{S}$ the closure (in the norm topology induced by $||\cdot||_U$) of S in $\hat{E}_U$. Let a > 0 be such that $V \subset aU$.

We will show that

$$d_n(V\cap F, U\cap F) = d_n(V\cap F, U\cap F; F) \leq d_n(V\cap F, \hat{U}\cap\hat{F}; \hat{F})$$

$$\leq d_n(V\cap F, \hat{U}; \hat{E}_U) \leq d_n(V,U;\hat{E}_U) = d_n(V,U) \qquad (4)$$

The first equality is (6.1.2) iv).

For the next inequality let $L \varepsilon \mathcal{F}_n(\hat{F})$ and $\delta > 0$ such that

$$V\cap F \subset \delta\hat{U}\cap\hat{F} + L \quad .$$

Then, since L is finite dimensional and $\hat{U}$ is $||\cdot||_{\hat{U}}$ bounded, then given $\eta > 0$ we can find $M \varepsilon \mathcal{F}_n(F)$ such that

$$z \varepsilon (a+\delta)\hat{U}\cap L \Rightarrow \exists\ z_1 \varepsilon M \ni ||z-z_1||_{\hat{U}} \leq \eta .$$

So let $x \varepsilon V\cap F$ which implies that $x = \delta y + z$, $y \varepsilon \hat{U}\cap\hat{F}$, $z \varepsilon L$. Hence $z \varepsilon (a+\delta)\hat{U}\cap L$ so we have $z_1 \varepsilon M$ with $z - z_1 \varepsilon \eta\hat{U}$. Thus,

$$x = (\delta+\eta)(\frac{\delta}{\delta+\eta}y + \frac{1}{\delta+\eta}(z-z_1)) + z_1 = (\delta+\eta)y_1+z_1 ,$$

and $x, z_1 \varepsilon F$ so $y_1 \varepsilon F$. Moreover,

$$y_1 = \frac{\delta}{\delta+\eta}y + \frac{1}{\delta+\eta}(z-z_1) \ \varepsilon \ \frac{\delta}{\delta+\eta}\hat{U} + \frac{\eta}{\delta+\eta}\hat{U} \subset \hat{U}$$

so $y_1 \varepsilon \hat{U} \cap F \subset \hat{U} \cap E = U$, since $U = \{x\varepsilon E: ||x||_U \leq 1\}$. Hence

$$x\varepsilon(\delta+\eta)U\cap F + M \quad .$$

Since this holds for any $\eta > 0$ the inequality is established.

For the next inequality in (4) we use the fact that $\hat{E}_U$ is a Hilbert space so we have the orthogonal projection $P: \hat{E}_U \to \hat{E}_U$ onto $\hat{F}$. If $L\varepsilon\mathcal{F}_n(\hat{E}_U)$ then $P(L)\varepsilon\mathcal{F}_n(\hat{F})$ and $P(\hat{U}) = \hat{U}\cap\hat{F}$. Hence, if, also, $V\cap F\subset\delta\hat{U} + L$ then

$$V\cap F = P(V\cap F)\subset\delta P(\hat{U}) + P(L) = \delta(\hat{U}\cap F) + P(L)$$

which establishes the inequality.

The next inequality in (4) follows from (6.1.2) i) and the last equality follows from (6.1.2) iv). Hence (4) holds and so the lemma is established. ∎

(6.2.3) Lemma

Let U,V be absolutely convex subsets of a vector space E such that U absorbs V. Let $g: E \to F$ be a linear map. Then,

$$d_n(g(V),g(U)) \leq d_n(V,U) \quad .$$

Proof

If $V\subset\delta U + L$ then $g(V)\subset\delta g(U) + g(L)$. ∎

(6.2.4) Theorem

Let E be a nuclear Fréchet space, $h: [0,\infty)\times \mathbb{N} \to [0,\infty)$ a function and $0 < r \leq \infty$. Consider the following two conditions for a locally convex space:

i) For each nbd of 0 U and $0 < \rho < r$ there is a nbd of 0 V such that

$$d_n(V,U) \leq h(\rho,n) \text{ for n sufficiently large,}$$

ii) For each nbd of 0 U there is a nbd of 0 V such that for each $\rho$ with $0 < \rho < r$ it follows that

$$d_n(V,U) \leq h(\rho,n) \text{ for n sufficiently large.}$$

Then, if E satisfies one of these two conditions and admits a continuous norm and F is a subspace of E then F satisfies the same condition. Moreover, if E satisfies one of these two conditions and F is a quotient space of E then F satisfies the same condition.

Proof

For the first statement we use the standard fact [53] that since E is nuclear and admits a continuous norm, then it has a fundamental system of nbds of 0 consisting of closed absolutely convex sets each of whose gage is a norm defined by an inner product. Then we apply (6.2.2).

For the second statement we apply (6.2.3) without any preliminaries. ∎

(6.3) Computation of Diameters

(6.3.1) A basis $(x_n)$ in a nuclear Fréchet space E is regular if its topology is defined by a sequence of norms $(||\cdot||_k)$ such that

$$\frac{||x_{n+1}||_k}{||x_{n+1}||_{k+1}} \leq \frac{||x_n||_k}{||x_n||_{k+1}} \quad \forall\ n, k .$$

In this case, it follows from the considerations of (3.2) that a fundamental sequence of nbds of 0 is given by $(V_k)_k$ where

$$V_k = \{x = \sum_n t_n x_n \varepsilon E: \sum_n |t_n|\ ||x_n||_k \leq 1\} .$$

It is then a standard result that the diameters, $d_n(V_j,V_k)$, $k < j$ are easy to compute.

(6.3.2) Proposition

In the context of (6.3.1) if $k < j$ we have

$$d_n(V_j,V_k) = \frac{||x_n||_k}{||x_n||_n} \quad \forall n .$$

(6.3.3) The importance of (6.3.2) lies in the easily checked fact that the coordinate basis is a regular basis in every power series space of infinite or finite type and in every $L_f(\alpha,r)$ space, $-\infty < r \leq \infty$. Hence we can compute the Kolmogorov diameters for these spaces without any difficulty.

(6.3.4) Another application of regular bases is to quasi-equivalence. A nuclear Fréchet space E with basis is said to have the quasi-equivalence property if for every two bases $(x_n)$ and $(y_n)$ for E there is a permutation $\pi$ of $\mathbb{N}$ and a sequence $(t_n)$ of positive scalars such that there is an isomorphism $T: E \to E$ with $Tx_n = t_n y_{\pi(n)}$.

(6.3.5) Theorem

Any nuclear Fréchet space with a regular basis has the quasi-equivalence pŕoperty. Moreover if $(x_n)$ and $(y_n)$ are two regular bases in a nuclear Fréchet space, then they satisfy the requirement of the quasi-equivalence property with $\pi$ equal to the identity map.

(6.3.6). A simple application of (6.3.3) and (6.3.5) which we will use is the fact that no two of the spaces $\Lambda_\infty(\alpha^1)$, $\Lambda_1(\alpha^2)$, $L_f(\alpha^3,-1)$, $L_f(\alpha^4,0)$, $L_f(\alpha^5,1)$, $L_f(\alpha^6,\infty)$ are isomorphic whatever the exponent sequences or the Dragilev functions.

## 7. Notes and Remarks

Our definition of complemented subspaces in (2.2) is not quite right. Usually it is explicitly required that the map $F\times G \to E$ by $(x,y) \rightsquigarrow x+y$ be an isomorphism. In all of our uses of this term, E will be a Fréchet space and in that case our definition is correct because of the open mapping theorem.

Nuclear Fréchet spaces were introduced by Grothendieck [37] who derived their elementary properties. An extensive study of nuclear Fréchet spaces in connection with bases and diameters was made by Mitiagin [45]. The proof of the isomorphism of K(a) and E in (3.2) is an easy consequence of the open mapping theorem and the absolute basis theorem of Dynin and Mitiagin [35]. The same is true of the statement in (3.4).

Power series spaces were explicitly introduced by Grothendieck [37]. For a detailed study of their properties see [26], [27] and [28] where also appear some early studies of subspaces.

The spaces $L_f(\alpha,r)$ were introduced by Dragilev [19] who also proved the statement in proposition (5.2), along with other properties of these spaces. The restrictions which we introduce in propositions (5.3) and (5.4) do not eliminate any particular space but they serve to greatly simplify some of our subsequent calculations. They appear here explicitly for the first time.

The Kolmogorov diameters were introduced in [40]. In addition to [45] mentioned above, their relation to nuclear Fréchet spaces was studied by Bessaga, Pełczyński and Rolewicz [11]. The facts in (6.1.2) are well known and their proofs are elementary. Lemma (6.2.2) is generally known. The use of (6.2.4) appears here for the first time.

Regular bases were introduced by Dragilev [19] and studied extensively there and in other papers. A proof of (6.3.2) can be found in [53].

Quasi-equivalence was first studied by Dragilev [18] and also Mitiagin [46]. Theorem (6.3.5) is due to Crone and Robinson [13].

## CHAPTER II

## GENERALITIES

1. Type $(d_i)$, $i = 0,1,\dots,6$

(1.1) Let $a = (a_n^k)$ be a matrix satisfying the conditions of equation (2) of I (3.2). We say that a is of type $(d_i)$, $i = 0,1,\dots,6$ if the one of the following conditions corresponding to i is satisfied.

$$\frac{a_{n+1}^k}{a_{n+1}^{k+1}} \le \frac{a_n^k}{a_n^{k+1}} \quad \forall\, n,k \tag{$d_0$}$$

$$\exists\, k \ni \forall j\ \exists \ell \ni \sup_n \frac{(a_n^j)^2}{a_n^k\, a_n^\ell} < \infty \tag{$d_1$}$$

$$\forall\, k\ \exists\, j \ni \forall \ell,\ \sup_n \frac{a_n^k\, a_n^\ell}{(a_n^j)^2} < \infty \tag{$d_2$}$$

$$\frac{a_n^{k+1}}{a_n^k} \le \frac{a_n^{k+2}}{a_n^{k+1}} \quad \forall\, n,k \tag{$d_3$}$$

$$\frac{a_n^{k+1}}{a_n^k} \ge \frac{a_n^{k+2}}{a_n^{k+1}} \quad \forall\, n,k \tag{$d_4$}$$

$$\exists\, M \ge 1 \ni \frac{a_n^{k+1}}{a_n^k} \le \left(\frac{a_n^{k+2}}{a_n^{k+1}}\right)^M \quad \forall\, n,k \tag{$d_5$}$$

$$\frac{a_n^{k+1}}{a_n^k} \ge \left(\frac{a_n^\ell}{a_n^{k+1}}\right)^M \quad \forall\, n,k,\ell \tag{$(d_6)_M$}$$

(Notice that $(d_6)_M$ depends on M).

(1.2) We can mention some easy examples of these types. Let $\alpha = (\alpha_n)$ be an exponent sequence of finite type.

(1.2.1) If $a_n^k = e^{k\alpha_n}$ then a is of type $(d_0)$, $(d_1)$, $(d_3)$, $(d_4)$ and $(d_5)$ but not $(d_2)$ or $(d_6)_M$.

(1.2.2) If $a_n^k = e^{-\frac{1}{k}\alpha_n}$ then a is of type $(d_0)$, $(d_2)$, $(d_4)$, $(d_5)$ and $(d_6)_{\frac{1}{2}}$ but not $(d_1)$ or $(d_3)$.

(1.2.3) If $a_n^k = e^{k\alpha_n}$ or $e^{-\frac{1}{k}\alpha_n}$ according as to whether n is even or odd then a is $(d_4)$ and $(d_5)$ but not $(d_0)$, $(d_1)$, $(d_2)$, $(d_3)$ or $(d_6)_M$.

(1.2.4) If $a_n^k = e^{-\frac{1}{2^k}\alpha_n}$ then a is $(d_5)$ but not $(d_4)$.

(1.2.5) The above examples show that no two of types $(d_i)$ can be equivalent with the exception of the pairs $(d_1)$, $(d_3)$ and $(d_2)$, $(d_6)_M$. These are also not equivalent because it is easy to violate $(d_3)$ and $(d_6)_M$ by changing $a_n^k$ for $k \geq n$. But any such change does not destroy the validity of $(d_1)$, $(d_2)$. This is a little artificial and we will see later that there is good reason for it.

(1.3) Now suppose that the matrix a is a representation of a basis $(x_n)$ in a nuclear Fréchet space E, corresponding to a fundamental sequence of seminorms $(\|\cdot\|_k)$. We consider what happens if we take another representation b of $(x_n)$ corresponding to an equivalent sequence $(|\cdot|_k)$.

(1.3.1) Clearly, if i = 1,2 then b is type $(d_i)$ iff a is, but this is not the case for any of the four types. Hence we say that $(x_n)$ is a <u>basis of type</u> $(d_i)$, i = 0,1,...,5 if it has a representation of this type. We say that it is <u>type</u> $(d_6)$ if $\forall M \geq 1$ it has a representation which is type $(d_6)_M$. We will need to know how to determine whether a basis is of a given type from a given sequence of seminorms.

(1.3.2) Proposition

If $(x_n)$ is a basis of type $(d_3)$ in a nuclear Fréchet space E and b is a representation of $(x_n)$ then there is a subsequence $(j_k)$ of indices and a sequence of positive numbers $(\varepsilon_k)$ $\ni$

$$\varepsilon_k \frac{b_n^{j_{k+1}}}{b_n^{j_k}} \leq \frac{b_n^{j_{k+2}}}{b_n^{j_{k+1}}} \quad \forall\ n,k$$

Proof

Let a be $(d_3)$ representation of $(x_n)$ and choose $j_1$ and $M_1 > 0 \ni a_n^1 \leq M_1 b_n^{j_1} \;\forall\ n$ and suppose that $j_1 < \ldots < j_{k+1}$ and $\varepsilon_1, \ldots, \varepsilon_{k-1}$ have been chosen as desired. We complete the induction by choosing $\ell < j$, $M,N,P > 0$ and $j_{k+2} > j_{k+1}$ $\ni$

$$a_n^{\ell} \leq M b_n^{j_k} \leq M b_n^{j_{k+1}} \leq N a_n^{j} \leq N a_n^{2j-\ell} \leq P b_n^{j_{k+2}} \quad \forall\ n$$

so

$$\frac{b_n^{j_{k+1}}}{b_n^{j_k}} \leq N \frac{a_n^{j}}{a_n^{\ell}} \leq N \frac{a_n^{2j-\ell}}{a_n^{j}} \leq \frac{NP}{M} \frac{b_n^{j_{k+2}}}{b_n^{j_{k+1}}} \quad \forall\ n$$

so we may set $\varepsilon_k = \frac{M}{NP}$. ∎

(1.3.3) Proposition

If $(x_n)$ is a basis of type $(d_5)$ in a nuclear Fréchet space E and b is a representation of $(x_n)$ then there is a subsequence $(j_k)$ of indices and a sequence $(M_k)$ of positive numbers $\ni$

$$\frac{b_n^{j_{k+1}}}{b_n^{j_k}} \leq \left(\frac{b_n^{j_{k+2}}}{b_n^{j_{k+1}}}\right)^{M_k} \quad \forall\, n,k$$

<u>Proof</u>

Let a be a $(d_5)$ representation. Using the nuclearity we can select by induction subsequences of indices $(j_k)$, $(\ell_k)$ ∋ for each k ∃ $n_k$ ∋

$$b_n^{j_{k+1}} \leq a_n^{\ell_{k+1}} \leq a_n^{2\ell_{k+1}-\ell_{k-1}} \leq b_n^{j_{k+2}} \quad \text{for } n \geq n_k \quad .$$

Then for such n we have

$$\frac{b_n^{j_{k+1}}}{b_n^{j_k}} \leq \frac{a_n^{\ell_{k+1}}}{a_n^{\ell_{k-1}}} \leq \left(\frac{a_n^{2\ell_{k+1}-\ell_{k-1}}}{a_n^{\ell_{k+1}}}\right)^{M^{\ell_{k+1}-\ell_{k-1}}} \leq \left(\frac{b_n^{j_{k+2}}}{b_n^{j_{k+1}}}\right)^{M^{\ell_{k+1}-\ell_{k-1}}}$$

and it suffices to take $M_k$ sufficiently larger than $M^{\ell_{k+1}-\ell_{k-1}}$ so that the inequality holds for all n. ∎

(1.3.4) At this point we can mention some gaps. We do not have converses for (1.3.2) and (1.3.3). Also, there should be similar results for $(d_4)$ and $(d_6)$. We do have something but the best formulation is not clear. For $(d_0)$ we have a much different situation which we discuss next. All of this could bear further investigation.

(1.3.5) A basis of type $(d_0)$ was called, in I (6.3.1) a <u>regular basis</u>. We say that a basis is <u>pseudo-regular</u> if it has a representation a ∋

$$\forall i \; \exists j \ni \forall k \; \exists \ell \ni \; \sup_{m \le n} \frac{a_n^i}{a_n^\ell} \, \frac{a_m^k}{a_m^j} < \infty \; .$$

It is quite clear that a regular basis is pseudo-regular and that this definition is independent of the representation. In many situations results about regular bases can be proved for pseudo-regular bases.

(1.3.6) It is an open question whether every pseudo-regular basis is regular.

(1.4) We now consider some relations between these types.

(1.4.1) Proposition

A basis is of type $(d_1)$ iff it is of type $(d_3)$.

Proof

It is immediate that any $(d_3)$ representation is also $(d_1)$. Conversely, if a basis has a $(d_1)$ representation then it has a representation b $\ni$

$$\sup_n \frac{(b_n^k)^2}{b_n^1 \, b_n^{k+1}} < \infty \quad \forall \, k .$$

Thus we have $M_k > 0 \ni$

$$\frac{b_n^{k+1}}{b_n^k} \le \frac{b_n^{k+1}}{b_n^1} \le M_k \, \frac{b_n^{k+2}}{b_n^{k+1}} \quad \forall \, n,k \quad .$$

Then if $(t_k)$ is a sequence of positive numbers satisfying

$$t_{k+2} \ge M_k \, \frac{t_{k+1}^2}{t_k} \; , \quad t_{k+1} \ge t_k$$

then $a_n^k = t_k\, b_n^k$ is also a representation of this basis and

$$\frac{a_n^{k+1}}{a_n^k}\,\frac{a_n^{k+1}}{a_n^{k+2}} = \frac{t_{k+1}^2}{t_k t_{k+2}}\,\frac{(b_n^{k+1})^2}{b_n^k\, b_n^{k+2}} \leq 1 .$$

∎

(1.4.2) Proposition

If E is a nuclear Fréchet space with a basis $(x_n)$ that is of type $(d_3)$ and also of type $(d_4)$, then E is isomorphic to an infinite type power series space.

Proof

The basis $(x_n)$ has a representation a of type $(d_4)$ and also satisfies the relation of (1.3.2). Thus we have

$$\frac{a_n^{j_{k+2}}}{a_n^{j_{k+1}}} \geq \varepsilon_1 \cdots \varepsilon_k \frac{a_n^{j_2}}{a_n^{j_1}} \geq (\varepsilon_1 \cdots \varepsilon_k) \frac{a_n^{j_1+1}}{a_n^{j_1}} \quad \forall\ n,k$$

so if we set $\beta_n = \dfrac{a_n^{j_1+1}}{a_n^{j_1}}$ we have $\forall$ n,k

$$a_n^{j_{k+2}} \geq (\varepsilon_1 \cdots \varepsilon_k) a_n^{j_{k+1}} \beta_n \geq \cdots \geq \varepsilon_1 (\varepsilon_1 \varepsilon_2) \cdots (\varepsilon_1 \cdots \varepsilon_k) a_n^{j_1} (\beta_n)^{k+1}$$

Hence $K(a) \subset K((a_n^{j_1} (\beta_n)^{k+1}))$.

On the other hand by $(d_4)$,

$$\frac{a_n^{k+2}}{a_n^{k+1}} \leq \frac{a_n^{j_1+1}}{a_n^{j_1}} = \beta_n \quad \text{for all n and } k \geq j_1 ,$$

so

$$a_n^{k+2} \leq a_n^{k+1}\beta_n \leq \dots \leq a_n^{j_1}(\beta_n)^{k+2-j_1}, \quad k \geq j_1$$

so $K(a) = K((a_n^{j_1}(\beta_n)^{k+1}))$. But the latter space is isomorphic (via a diagonal transformation) to $\Lambda_\infty(\alpha)$, $\alpha_n = \log\beta_n$. ∎

(1.4.3) Proposition

If E is a nuclear Fréchet space with a basis $(x_n)$ that is of type $(d_2)$ and also of type $(d_5)$ then E is isomorphic to a finite type power series space.

Proof

Let b be a $(d_2)$ representation for $(x_n)$ and define the matrix a by

$$a_n^k = \begin{cases} \dfrac{b_n^k}{b_n^n} & k \leq n \\ 1 & k > n \end{cases}$$

Clearly a is a representation of a diagonal transform $(y_n)$ of $(x_n)$ and we may assume (by passing to a subsequence on k) that

$$a_n^k \leq (a_n^{k+1})^2 \quad n \text{ sufficiently large}$$

and we may adjust the matrix so that this holds $\forall n,k$ and still $a_n^k \leq a_n^{k+1} \leq 1$. Hence we have

$$a_n^k a_n^\ell \leq (a_n^{k+1})^2 \quad \forall n,k ,$$

and this continues to hold if k runs through a subsequence. Hence by (1.3.3) we may assume that this relation holds and $\exists\ M_k \ni$

$$\frac{a_n^{k+1}}{a_n^k} \leq \left(\frac{a_n^{k+2}}{a_n^{k+1}}\right)^{M_k} \quad \forall\ n,k\ .$$

Thus we can find $M_{nk}$ with $0 < M_{nk} \leq M_k \ni$

$$\frac{a_n^{\ell}}{a_n^{k+1}} \leq \frac{a_n^{k+1}}{a_n^k} = \left(\frac{a_n^{k+2}}{a_n^{k+1}}\right)^{M_{nk}} \quad \forall\ n,k,\ell \quad .$$

From which it follows that $a_n^k = a_n^2 (\beta_n)^{F_{nk}}$ $\forall$ n and $k > 2$ where

$$\beta_n = \frac{a_n^2}{a_n^1} > 1 \text{ and } F_{nk} = \frac{1}{M_{n1}} + \frac{1}{M_{n1}M_{n2}} + \ldots + \frac{1}{M_{n1}\cdots M_{n,k-2}}\ .$$

Since

$$\frac{a_n^{k+1}}{a_n^k} = (\beta_n)^{F_{n,k+1}-F_{nk}},\quad \frac{a_n^{\ell}}{a_n^{k+1}} = (\beta_n)^{F_{n\ell}-F_{n,k+1}},\quad \ell > k+1$$

it follows that $F_{n\ell}-F_{n,k+1} \leq F_{n,k+1}-F_{nk}$ so $F_n = \sup_k F_{nk} < \infty$ . We then write

$$\gamma_n = \beta_n^{F_n},\ s_{nk} = 1 - \frac{F_{nk}}{F_n} \text{ so that } a_n^k = a_n^2\, \gamma_n \gamma_n^{-s_{nk}}\ .$$

We will complete the proof by showing that $K(\gamma_n^{-s_{nk}}) = \Lambda_1(\log\gamma_n^{s_{n3}})$.

First we show that for $k \geq 3$, $\inf_n \frac{s_{n,k+1}}{s_{nk}} > 0$. If this were not so for some k then we would have an infinite set $\mathbb{N}_0 \subset \mathbb{N}$ with $\lim_{n \in \mathbb{N}_0} \frac{s_{n,k+1}}{s_{nk}} = 0$. But then

$$0 < \frac{1}{M_k} \leq \frac{1}{M_{nk}} = \frac{F_{n,k+2}-F_{n,k+1}}{F_{n,k+1}-F_{n,k}} = \frac{s_{n,k+1}-s_{n,k+2}}{s_{nk}-s_{n,k+1}} \leq \frac{s_{n,k+1}}{s_{nk}-s_{n,k+1}} =$$

$$= \frac{s_{n,k+1}}{s_{nk}} \frac{1}{1-\frac{s_{n,k+1}}{s_{nk}}}$$

which is a contradiction.

Now $F_{nk}$ is increasing in k so $\lim_k s_{nk} = 0$ and

$s_{n,k+1} - s_{n\ell} \leq s_{nk} - s_{n,k+1}$ $\forall$ n and $k,\ell \geq 3$. Hence $s_{n,k+1} \leq \frac{1}{2}s_{nk}$ so $\lim_k \sup_n \frac{s_{nk}}{s_{n3}} \leq \lim_k 2^{-k+3} = 0$.

Thus, on the one hand, if we are given $k > 3$, we can find $j \ni \frac{s_{nk}}{s_{n3}} > \frac{1}{j}$ $\forall$ n so

$$\gamma_n^{-s_{nk}} \leq \left(\gamma_n^{-s_{n3}}\right)^{\frac{1}{j}} = e^{-\frac{1}{j}\log\gamma_n^{s_{n3}}} \quad \forall\, n$$

and on the other hand given j we have $\ell \ni \frac{s_{n\ell}}{s_{n3}} \leq \frac{1}{j}$ $\forall$ n so

$$e^{-\frac{1}{j}\log\gamma_n^{s_{n3}}} = \left(\gamma_n^{-s_{n3}}\right)^{\frac{1}{j}} \leq \gamma_n^{-s_{n\ell}} \quad \forall\, n$$

∎

(1.4.4) Proposition

Any basis of type $(d_6)$ in a nuclear Fréchet space is also type $(d_2)$.

Proof

Let a be a $(d_6)_M$ representation of the basis and, given k to verify the $(d_2)$ condition, take j = k+1. Then, since $M \geq 1$ we have, for $\ell \geq k+1$,

$$\frac{a_n^k a_n^\ell}{(a_n^{k+1})^2} \leq \left(\frac{a_n^{k+1}}{a_n^\ell}\right)^{M-1} \leq 1 \quad \forall\, n \quad .$$

but the inequality is obvious for $\ell < k+1$. ∎

(1.4.5) Corollary

If E is a nuclear Fréchet space with a basis $(x_n)$ that is of type $(d_6)$ and also of type $(d_5)$ then E is isomorphic to a finite type power series space.

Proof

Immediate from (1.4.3) and (1.4.4).

(1.4.6) Proposition

Let a be a matrix of type $(d_5)$. Then either a is type $(d_1)$ or there is a subsequence of indices $\mathbb{N}_1 \ni$ the stepspace $(K(a))_{\mathbb{N}_1}$ is a permutation and diagonal transform of a finite type power series space.

Proof

Suppose that a is not $(d_1)$ so in particular $\exists\, j \ni \forall\, \ell\ \exists$ subsequence $\mathbb{N}_\ell$ of $\mathbb{N} \ni$

$$\lim_{n \in \mathbb{N}_\ell} \frac{(a_n^j)^2}{a_n^1 a_n^\ell} = \infty \ .$$

Since $a_n^k$ increases in k we can apply a diagonal argument to show that $\mathbb{N}_\ell$ is independent of $\ell$. That is, we have j and $\mathbb{N}_1 \ni$

$$\lim_{n\in\mathbb{N}_1} \frac{(a_n^j)^2}{a_n^1 a_n^\ell} = \infty \quad \forall \ell .$$

Also, this relation remains true if j is increased.

Now we will show that the submatrix $(a_n^k)_{n\in\mathbb{N}_1}$ is not $(d_1)$. If it were, then applying (1.4.1) and (1.3.2) we would have a subsequence of indices $(i_k)$ with $i_2 = j$ (increasing j if necessary) and $i_3 \geq 2j-1 \ni$

$$\sup_{n\in\mathbb{N}_1} \frac{(a_n^{i_{k+1}})^2}{a_n^{i_k} a_n^{i_{k+2}}} < \infty \quad .$$

Now $1 < j = i_2 < 2j-1 \leq i_3$ so by the $(d_5)$ condition and this last relation we can find indices $k,\ell$ and $C > 0 \ni$

$$\frac{a_n^j}{a_n^1} = \frac{a_n^j}{a_n^{j-1}} \cdots \frac{a_n^2}{a_n^1} \leq \left(\frac{a_n^{2j-1}}{a_n^{2j-2}} \cdots \frac{a_n^{j+1}}{a_n^j}\right)^{M^{j-1}} = \left(\frac{a_n^{2j-1}}{a_n}\right)^{M^{j-1}} \leq$$

$$\left(\frac{a_n^{i_3}}{a_n^j}\right)^{M^{j-1}} = \underbrace{\frac{a_n^{i_3}}{a_n^{i_2}} \cdots \frac{a_n^{i_3}}{a_n^{i_2}}}_{M^{j-1} \text{ copies}} \leq C\, \frac{a_n^{i_3}}{a_n^{i_2}} \frac{a_n^{i_4}}{a_n^{i_3}} \cdots \frac{a_n^\ell}{a_n^k} = C\, \frac{a_n^\ell}{a_n^{i_2}} = C\, \frac{a_n^\ell}{a_n^j}$$

which contradicts the fact that the above limit is $\infty$. Thus $a|_{\mathbb{N}_1}$ is not $(d_1)$.

We repeat this argument to obtain subsequences $\mathbb{N}_1 \supset \mathbb{N}_2 \supset \cdots \ni$ for each k we have $j_k \ni$

$$\lim_{n\in \mathbb{N}_k} \frac{(a_n^{j_k})^2}{a_n^k a_n^\ell} = \infty \quad \forall \ell .$$

But then we can find a single subsequence $\mathbb{N}_\infty \ni \mathbb{N}_\infty \sim \mathbb{N}_k$ is finite so it follows that $\forall k \; \exists \; j \ni$

$$\lim_{n\in \mathbb{N}_\infty} \frac{a_n^j}{a_n^k a_n^\ell} = \infty \quad \forall \ell$$

That is, $a|_{\mathbb{N}_\infty}$ is type $(d_2)$ and the result follows from (1.4.3) ∎

(1.4.7) All of the results in this section arose from a specific need which we will describe later. No doubt there are other relations and other possible applications. Also, there is no clear pattern. It is possible that some investigation here might be fruitful.

At the very least, one might be able to develop a more reasonable notation.

(1.5) We give now some additional lemmas which will be required.

(1.5.1) Lemma

If a is a matrix of type $(d_3)$ and $A > 0$, then $\exists$ a subsequence $(j_k)$ of indices $\ni j_1 = 1$ and

$$\left(\frac{a_n^{j_{k+1}}}{a_n^{j_k}}\right)^A \leq \frac{a_n^{j_{k+2}}}{a_n^{j_{k+1}}} \quad \forall \; n,k .$$

<u>Proof</u>

$$\left(\frac{a_n^{2^k}}{a_n^{2^{k-1}}}\right)^2 \leq \frac{a_n^{2^{k+1}}}{a_n^{2^k+2^{k-1}}} \; \frac{a_n^{2^k+2^{k-1}}}{a_n^{2^k}} = \frac{a_n^{2^{k+1}}}{a_n^{2^k}} ,$$

so if $A \leq 2$ we may take $j_k = 2^{k-1}$. If $A > 2$ we merely repeat this process an appropriate number of times. ∎

(1.5.2) Lemma

If a is a matrix of type $(d_6)_M$ and $(j_k)$ is any subsequence of indices, then $(a_n^{j_k})$ is again a matrix of type $(d_6)_M$.

Proof

Applying the fact that $(a_n^k)$ is increasing in k and the $(d_6)_M$ condition we have,

$$\frac{a_n^{j_{k+1}}}{a_n^{j_k}} \geq \frac{a_n^{j_{k+1}}}{a_n^{j_{k+1}-1}} \geq \left(\frac{a_n^{j_\ell}}{a_n^{j_{k+1}}}\right)^M \quad \forall\, n,k,\ell$$ ∎

## 2. The Basic Construction Tools

(2.1) In this section we develop the basic tools which will be used in all of our constructions of subspaces and quotient spaces. The fundamental idea here is that because of the nuclearity, a fundamental sequence of seminorms for a Köthe space K(a) is given by $(|\cdot|_k)$ where

$$|\xi|_k = \sup_n |\xi|_n\, a_n^k \quad .$$

Again because of the nuclearity, $\lim_n |\xi_n| a_n^k = 0 \;\forall\, k$ so that the above sup actually occurs at some index n. The essential point of our investigation is to study how this index at which the sup occurs can vary as all the other parameters such as $\xi$, k, a are varied.

This approach is used for subspaces. In quotient spaces, we study a dual relation.

(2.2) We begin with subspaces.

(2.2.1) Let $a = (a_n^k)$ be an infinite matrix of positive numbers. We say that a is of <u>strict type</u> $(d_0)$ if

$$\frac{a_n^{k+1}}{a_{n+1}^{k+1}} < \frac{a_n^k}{a_{n+1}^k} \quad \forall\ n,k \ .$$

Relative to this matrix, if $t_1,\ldots,t_r$ are scalars not all 0 and $k \in \mathbb{N}$ we define

$$q^k(t_1,\ldots,t_r) = \max\{q : \max_{1 \le n \le r} |t_n| a_n^k = |t_q| a_q^k\} \ .$$

(2.2.2) Proposition

In the context of (2.2.1), $q^k(t_1,\ldots,t_r) \le q^{k+1}(t_1,\ldots,t_r)$.

<u>Proof</u>

Let $q = q^k(t_1,\ldots,t_r)$ and $n < q$. Then $|t_n| a_n^k \le |t_q| a_q^k$ so by the $(d_0)$ condition,

$$\frac{|t_n|}{|t_q|} \le \frac{a_q^k}{a_n^k} < \frac{a_q^{k+1}}{a_n^{k+1}} \ ,$$

so $|t_n| a_n^{k+1} < |t_q| a_q^{k+1}$ from which the result follows. ∎

(2.2.3) Proposition

If the context of (2.2.1) let $0 = j_0 < j_1 < \ldots < j_m$ be integers with $m \le r$ and let $q^1 < q^2 < \ldots < q^m$ be positive integers with $q^m \le r$. Then ∃ scalars $t_1,\ldots,t_r$ ∋

$$q^k(t_1,\ldots,t_r) = q^\ell \text{ for } j_{\ell-1} < k \le j_\ell,\ \ell = 1,\ldots,m \ .$$

Specifically, this holds for any choice of $t_1,\ldots,t_r \ni t_{q^1} \neq 0$ but is otherwise arbitrary, $t_\ell = 0$ for $\ell \neq q^1,\ldots,q^m$ and

$$\frac{a_{q^\ell}^{j_\ell+1}}{a_{q^{\ell+1}}^{j_\ell+1}} \leq \frac{|t_{q^{\ell+1}}|}{|t_{q^\ell}|} < \frac{a_{q^\ell}^{j_\ell}}{a_{q^{\ell+1}}^{j_\ell}} \qquad \ell = 1,\ldots,m-1 \quad .$$

Proof

Such a choice of $t_1,\ldots,t_r$ is possible because of the strict $(d_0)$ condition and the fact that $q^\ell < q^{\ell+1}$, so we may suppose it has been made. Clearly, $q^k(t_1,\ldots,t_r)$ can only be one of $q^1,\ldots,q^m$.

Now, applying the $(d_0)$ condition to the inequality determining the choice of $t_1,\ldots,t_r$ we have for $\ell = 1,\ldots,m-1$ and $k \leq j_\ell$,

$$\frac{|t_{q^{\ell+1}}|}{|t_{q^\ell}|} < \frac{a_{q^\ell}^k}{a_{q^{\ell+1}}^k} \quad \text{so} \quad |t_{q^{\ell+1}}|a_{q^{\ell+1}}^k < |t_{q^\ell}|a_{q^\ell}^k \quad .$$

Now the last inequality holds provided $1 \leq \ell \leq m-1$ and $k \leq j_\ell$. But $k \leq j_\ell$ remains true if $\ell$ is increased so we may repeat the argument for $\ell$ replaced by $\ell+1, \ell+2,\ldots,m-1$ and we have,

$$|t_{q^i}|a_{q^i}^k < |t_{q^\ell}|a_{q^\ell}^k \quad \text{for } \ell < i \leq m \text{ and } k \leq j_\ell \; .$$

This implies that $q^k(t_1,\ldots,t_r) \leq q^\ell$ for $k \leq j_\ell$, $\ell = 1,\ldots,m-1$.

Similarly, if $\ell = 2,\ldots,m$ and $k > j_{\ell-1}$,

$$\frac{a^k_{q^{\ell-1}}}{a^k_{q^\ell}} \leq \frac{|t_{q^\ell}|}{|t_{q^{\ell+1}}|} \quad \text{so} \quad |t_{q^{\ell-1}}|a^k_{q^{\ell-1}} \leq |t_{q^\ell}|a^k_{q^\ell}$$

and this time we replace $\ell$ by $\ell-1,\dots,2$ and repeat the argument to obtain,

$$|t_{q^i}|a^k_{q^i} \leq |t_{q^\ell}|a^k_{q^\ell} \text{ for } 1 \leq i < \ell \text{ and } k > j_{\ell-1} .$$

This implies that $q^k(t_1,\dots,t_r) \geq q^\ell$ for $k > j_{\ell-1}$, $\ell = 2,\dots,m$.

Hence if $\ell = 2,\dots,m-1$ we have our result. If $\ell = 1$ we apply only the first half and if $\ell = m$ only the second half.

∎

(2.2.4) The results of (2.2.2) and (2.2.3) describe completely what behavior of $q^k(t_1,\dots,t_r)$ is possible and also provide an explicit formula for realizing any possible behavior. This will be an important idea in our constructions.

(2.3) We have analogous results for quotient spaces.

(2.3.1) In the context of (2.2.1) we define

$$p^k(t_1,\dots,t_r) = \min\{p: \min_{1\leq n\leq r} \frac{a^k_n}{|t_n|} = \frac{a^k_p}{|t_p|}\} .$$

(2.3.2) Proposition

In the context of (2.3.1), $p^{k+1}(t_1,\dots,t_r) \leq p^k(t_1,\dots,t_r)$

Proof

Let $p = p^k(t_1,\dots,t_r)$ and $n > p$. Then $\dfrac{a^k_p}{|t_p|} \leq \dfrac{a^k_n}{|t_n|}$ so by the $(d_0)$ condition,

$$\frac{|t_n|}{|t_p|} \leq \frac{a_n^k}{a_p^k} < \frac{a_n^{k+1}}{a_p^{k+1}}$$

so

$$\frac{a_p^{k+1}}{|t_p|} < \frac{a_n^{k+1}}{|t_n|}$$

so $p^{k+1}(t_1,\ldots,t_r) \leq p$. ■

(2.3.3) Proposition

In the context of (2.3.1) let $0 = j_0 < j_1 < \ldots < j_m$ be integers with $m \leq r$ and let $p^1 > p^2 > \ldots > p^m$ be positive integers with $p^1 \leq r$.

Then $\exists$ scalars $t_1,\ldots,t_r \ni$

$$p^k(t_1,\ldots,t_r) = p^\ell \text{ for } j_{\ell-1} < k \leq \rho_\ell, \quad \ell=1,\ldots,m .$$

Specifically, this holds for any choice of $(t_1,\ldots,t_r) \ni t_{p^1} \neq 0$ but is otherwise arbitrary, $t_\ell = 0$ for $\ell \neq p^1,\ldots,p^m$ and

$$\frac{a^{j_\ell+1}_{p^{\ell+1}}}{a^{j_\ell+1}_{p^\ell}} < \frac{|t_{p^{\ell+1}}|}{|t_{p^\ell}|} \leq \frac{a^{j_\ell}_{p^{\ell+1}}}{a^{j_\ell}_{p_\ell}}, \quad \ell = 1,\ldots,m-1$$

<u>Proof</u>

Such a choice of $t_1,\ldots,t_r$ is possible because of the strict $(d_0)$ condition and the fact that $p^{\ell+1} < p^\ell$, so we may suppose it has been made. Clearly, $p^k(t_1,\ldots,t_r)$ can only be one of $p^1,\ldots,p^m$.

Now applying the $(d_0)$ condition to the inequality determining the choice of $t_1,\ldots,t_r$ we have, for $\ell = 1,\ldots,m-1$ and $k \le j_\ell$

$$\frac{|t_{p^{\ell+1}}|}{|t_{p^\ell}|} \le \frac{a^k_{p^{\ell+1}}}{a^k_{p^\ell}} \quad \text{so} \quad \frac{a^k_{p^\ell}}{|t_{p^\ell}|} \le \frac{a^k_{p^{\ell+1}}}{|t_{p^{\ell+1}}|} \quad .$$

As in the proof of (2.2.3) we may repeat this argument with $\ell$ replaced by $\ell+1,\ldots,m-1$ because $k \le j_\ell$ remains true if $\ell$ is increased so we may conclude that

$$\frac{a^k_{p^\ell}}{|t_{p^\ell}|} \le \frac{a^k_{p^i}}{|t_{p^i}|} \quad \text{for } \ell < i \le m \text{ and } k \le j_\ell \ .$$

This implies that $p^k(t_1,\ldots,t_r) \le p^\ell$ for $k \le j_\ell$, $\ell = 1,\ldots,m-1$.

Similarly, if $\ell = 2,\ldots,m$ and $k > j_{\ell-1}$,

$$\frac{a^k_{p^\ell}}{a^k_{p^{\ell-1}}} < \frac{|t_{p^\ell}|}{|t_{p^{\ell-1}}|} \quad \text{so} \quad \frac{a^k_{p^\ell}}{|t_{p^\ell}|} < \frac{a^k_{p^{\ell-1}}}{|t_{p^{\ell-1}}|} \ ,$$

and this time we replace $\ell$ by $\ell-1,\ldots,2$ and repeat the argument to obtain

$$\frac{a^k_{p^\ell}}{|t_{p^\ell}|} < \frac{a^k_{p^i}}{|t_{p^i}|} \quad \text{for } 1 \le i < \ell \text{ and } k > j_{\ell-1} \ .$$

This implies that $p^k(t_1,\ldots,t_r) \ge p_\ell$ for $k > j_{\ell-1}$, $\ell = 2,\ldots,m$.

Hence if $\ell = 2,\ldots,m-1$ we have our result. If $\ell = 1$ we apply only the first half and if $\ell = m$ only the second half. ∎

(2.3.4) The results of (2.3.2) and (2.3.3) describe completely what behavior of $p^k(t_1,\ldots,t_r)$ is possible and also provide an explicit formula for realizing any possible behavior. This will be an important idea in our constructions.

(2.4) In most of the applications of (2.2.3) and (2.3.3) we will be in the simpler case in which $j_\ell = \ell$. Also, it will generally be a situation in which $m = r$ and $q^\ell = \ell$ for subspaces while $p^\ell = m - \ell + 1$ for quotient spaces. In such situations the notations for our relations will be a little simpler.

## 3. Some First Constructions

Before turning to our main study of subspaces and quotient spaces, we show how the results of section 2 can be used to prove some interesting and perhaps surprising general properties of subspaces and quotient spaces of nuclear Fréchet spaces.

(3.1) Regular Bases

(3.1.1) Proposition

Every basis $(x_n)$ in a nuclear Fréchet space E not isomorphic to $\omega$ has a subsequence $(x_{j_n})$ which is a regular basis for the subspace F which it generates. Moreover, $(x_{j_n})$ has a representation which is of strict type $(d_0)$ and F is complemented in E.

Proof

By the nuclearity we can find a representation a for $x_n \ni \lim_n \frac{a_n^k}{a_n^{k+1}} = 0 \ \forall\, k$. Since E is not isomorphic to $\omega$ we can find a subsequence of indices $(j_n) \ni a_{j_n}^k > 0 \ \forall\, j,n$ and

$$\frac{a^k_{j_{n+1}}}{a^{k+1}_{j_{n+1}}} < \frac{a^k_{j_n}}{a^{k+1}_{j_n}} \quad \text{for } k = 1,2,\ldots,n-1 \quad .$$

Thus $(x_{j_n})$ has a representation, $(a^k_{j_n})$ ∋ the strict $(d_0)$ condition holds whenever $k < n$. Now we can easily change the values of $(a^k_{j_n})$ for $n \le k$ so that the strict $(d_0)$ condition holds $\forall\, n, k$. But it follows from the definition of Köthe space that since, for each k we are changing only finitely many n, $K((a^k_{j_n}))$ is not changed so $(x_{j_n})$ has a strict $(d_0)$ representation.

The fact that F is complemented is an immediate consequence of the absolute basis theorem of Dynin and Mitiagin [35]. ∎

(3.1.2) Corollary

Every nuclear Fréchet space has a subspace with a regular basis.

Proof

This follows from (3.1.1) and the well-known fact [9] that every Fréchet space has a subspace with a basis.

∎

(3.1.3) Proposition

Every nuclear Fréchet space has a quotient space with a regular basis. If the space is not isomorphic to $\omega$, then the quotient space can be chosen so as to admit a continuous norm.

Proof

If the space $E \cong \omega$ the result is trivial. If $E \not\cong \omega$ it has a fundamental sequence of barreled nbds of 0 $(U_k)$ ∋ each of the Banach spaces $E'_{U^0_k}$ (=subspace of E' generated by $U^0_k$ with norm given by the gage of $U^0_k$) is infinite dimensional. Let $||\cdot||_k$ be the norm in

$E'_{U_k^0}$ .

Fix $\varepsilon_n \in (0,1) \ni \prod_n (1-\varepsilon_n) > 0$. We will choose $(u_n)$ in $U_1^0 \ni$

$$||t_1u_1+\ldots+t_nu_n||_k \le \frac{1}{1-\varepsilon_n}||t_1u_1+\ldots+t_{n+1}u_{n+1}||_k \quad \forall\, k \le n$$

where $t_1,\ldots,t_{n+1}$ are arbitrary scalars.

Assume $u_1,\ldots,u_n$ have been chosen and let R be the space generated by $u_1,\ldots,u_n$ and $S_k = \{u\varepsilon R: ||u||_k=1\}$, $k = 1,\ldots,n$. Since $S_k$ is compact in the $||\cdot||_k$-topology, we can find an $\varepsilon_n$-net $v_1^k,\ldots,v_{q^k}^k$ in $S_k$ and by the Hahn-Banach theorem we can find linear functionals $\phi_1^k,\ldots,\phi_{q^k}^k$ on $E'_{U_1^0} \ni$

$$\phi_m^k(v_m^k) = 1, \quad \sup_{\substack{U\varepsilon E'_{U_1^0} \\ ||U||_k\le 1}} |\phi_m^k(u)| = 1, \quad 1 \le m \le q^k, \quad k \le n \; .$$

Since $E'_{U_1^0}$ is infinite dimensional we can choose $u_{n+1}\varepsilon U_1^0, u_{n+1}\neq 0$ and $u_{n+1} \varepsilon \bigcap_{m,k} \mathrm{KER}\, \phi_m^k$. Then for any m,k and scalar t we have

$$1 = \phi_m^k(v_m^k) = \phi_m^k(v_m^k+tu_{n+1}) \le ||v_m^k + tu_{n+1}||_k$$

so if $u \varepsilon S_k$ we can find $k,m \ni ||v_m^k - u|| \le \varepsilon_n$ so

$$||u+tu_{n+1}||_k \ge ||v_m^k+tu_{n+1}||_k - ||u-v_m^k|| \ge 1 - \varepsilon_n \; .$$

Thus $(u_n)$ has been chosen.

It then follows that $(u_n)$ is a basic sequence in each $E'_{U_k^0}$ so if $(x_n)$ is a sequence in E biorthogonal to $(u_n)$ it follows that $(x_n)$ is a basic sequence in each $(E'_{U_k^0})' = E''_{U_k} = \hat{E}_{U_k}$. Hence $(x_n)$ is a basic sequence in E.

Finally, since $u_n \in U_1^0$ and $\langle x_n, u_n \rangle = 1$ it follows that $\forall n$, $x_n \notin U_1$ from which it easily follows that the space generated by $(x_n)$ admits a continuous norm. Finally we apply (3.1.1). ∎

(3.1.4) Proposition

A regular basis in a nuclear Fréchet space with a continuous norm has a representation of strict type $(d_0)$.

<u>Proof</u>

Because of the continuous norm the basis has a representation a of type $(d_0)$ with $a_n^k > 0 \ \forall n,k$. We can find a sequence $(r_n^1)$ with $1 < r_n^1 < 2 \ni \dfrac{r_n^1 \ a_n^1}{a_n^2}$ is strictly decreasing in n. Then we can find a sequence $(r_n^2)$ with $1 < r_n < 2 \ni$

$$\frac{r_n^1 \ a_n^1}{r_n^2 \ a_n^2} \text{ and } \frac{r_n^2 \ a_n^2}{a_n^3} \text{ are strictly decreasing in n .}$$

This is done simply choosing $r_n^2$ sufficiently close to 1. This process is continued indefinitely and we obtain that $(r_n^k \ a_n^k)$ is a strictly $(d_0)$ representation of this basis. ∎

(3.2) Theorem

If E,F are two nuclear Fréchet spaces neither of which is

isomorphic to $\omega$, then there is a nuclear Fréchet space G isomorphic to a subspace of E and to a subspace of F.

Proof

A space not isomorphic to $\omega$ has a subspace with continuous norm [9] to which we can apply (3.1.2) and then (3.1.4). Thus we may assume that E,F have bases $(x_n)$, $(y_n)$ respectively with strict type $(d_0)$ representations a,b respectively. We also can arrange that $\lim_n \frac{a_n^k}{a_n^{k+1}} = \lim_n \frac{b_n^k}{b_n^{k+1}} = 0$. Decompose $(x_n)$, $(y_n)$ into $(x_{m,\nu})$ $(y_{m,\nu})$ $m,\nu \in \mathbb{N}$ so that for each m, $(a^k_{m,\nu})_{k,\nu}$ and $(b^k_{m,\nu})_{k,\nu}$ are still strict type $(d_0)$.

Fix m. We will select $q^1 < q^2 < \ldots < q^m$ and then $t_1,\ldots,t_{q^m}$ according to (2.2.3) relative to the matrix $(a^k_{m,\nu})_{k,\nu}$ and define $z_m \in E$ by

$$z_m = \sum_{i=1}^{m} t_{q^i} x_{m,q^i} .$$

We will also select $\tilde{q}^1 < \tilde{q}^2 < \ldots < \tilde{q}^m$ and then $\tilde{t}_1,\ldots,\tilde{t}_{\tilde{q}^m}$ according to (2.2.3) relative to the matrix $(b^k_{m,\nu})_{k,\nu}$ and define $\tilde{z}_m \in F$ by

$$\tilde{z}_m = \sum_{i=1}^{m} \tilde{t}_{\tilde{q}^i} y_{m,\tilde{q}^i} .$$

Then using the "sup" norms in E,F we have from (2.2.3),

$$|z_m|_k = |t_{q^k}| a^k_{m,q^k}, \quad |\tilde{z}_m|_k = |\tilde{t}_{\tilde{q}^k}| a^k_{m,\tilde{q}^k} \qquad k = 1,\ldots,m .$$

Moreover, $(z_m)$ and $(\tilde{z}_m)$ are block basic sequences so they are bases generating subspaces $G,\tilde{G}$ of E,F respectively.

It remains to select $q^k, \tilde{q}^k$. We have, from the inequality in (2.2.3), for $k = 1,\dots,m$

$$|z_m|_k \frac{a^{k+1}_{q^k}}{a^k_{q^k}} \le |z_m|_{k+1} < |z_m|_k \frac{a^{k+1}_{q^{k+1}}}{a^k_{q^{k+1}}} ,$$

$$|\tilde{z}_m|_k \frac{b^{k+1}_{\tilde{q}^k}}{b^k_{\tilde{q}^k}} \le |\tilde{z}_m|_{k+1} < |\tilde{z}_m|_k \frac{b^{k+1}_{\tilde{q}^{k+1}}}{b^k_{\tilde{q}^k}} .$$

Moreover, by an appropriate choice of $t_{q^{k+1}}$, $\tilde{t}_{\tilde{q}^{k+1}}$ we can have $|z_m|_k$, $|\tilde{z}_m|_k$ arbitrarily close to the right hand side of the inequality.

Thus we may select $q^1, \tilde{q}^1, q^2, \tilde{q}^2,\dots,q^m, \tilde{q}^m$ so that $|z_m|_k \le |\tilde{z}_m|_k \le |z_m|_{k+1}$ for $m \ge k$. Hence $G \cong \tilde{G}$. ∎

(3.3) For the common quotient spaces, we have to work a little harder, mainly to obtain the analog of block basic sequence.

(3.3.1) Proposition

Let E be a nuclear Fréchet space with basis $(x_n)$ and representation a. Let $0 = r_0 < r_{n-1} < r_n$ be integers and $(t_i)$ scalars $\ni \forall n \; \exists \; i(n) \in (r_{n-1}, r_n] \ni t_{i(n)} \neq 0$. Let $p_n^k = p^k(t_{r_{n-1}+1},\dots,t_{r_n})$ corresponding to a and let $b_n^k = \dfrac{a^k_{p_n^k}}{|t_{p_n^k}|}$. Then there is a quotient map $T: E \to K(b) \ni$

$$Tx_i = t_i e_n \text{ for } i \in (r_{n-1}, r_n],\ n \in \mathbb{N} \tag{1}$$

Conversely if $\tilde{K}$ is a sequence space with a locally convex topology $\ni$ the coordinate functionals are continuous and $T: E \to \tilde{K}$ is a quotient map satisfying (1) then $(e_n)$ is a basis for $\tilde{K}$ and $\tilde{K} = K(b)$.

Proof

We prove the second statement first. If $y \in \tilde{K}$, $\exists\, \xi \in E \ni$

$$y = T(\xi) \text{ so } \xi = \sum_i \xi_i x_i \text{ and } y = \sum_i \xi_i T x_i = \sum_n \Big( \sum_{i=r_{n-1}+1}^{r_n} \xi_i t_i \Big) e_n \text{ and since}$$

the coordinate functionals are continuous, the representation is unique so $(e_n)$ is a basis for $\tilde{K}$.

Since T is a quotient map the topology of $\tilde{K}$ is determined by the quotient norms $(||\cdot||_k)$ given by $||y||_k = \inf\{|x|_k : y=Tx\}$, $y \in K$ where

$$|x|_k = \Big(\sum_i (\xi_i a_i^k)^2\Big)^{\frac{1}{2}}, \quad x = \sum_i \xi_i x_i \in E .$$

Hence, for $n,k \in \mathbb{N}$

$$||e_n||_k^2 = \inf\Big\{\sum_m \sum_{i=r_{m-1}+1}^{r_m} (\xi_i a_i^k)^2 : \sum_i \xi_i x_i \in E \text{ and } \sum_{i=r_{m-1}+1}^{r_m} \xi_i t_i = \delta_{mn}\Big\}$$

$$= \inf\Big\{ \sum_{i=r_{m-1}+1}^{r_n} (\xi_i a_i^k)^2 : \sum_{i=r_{n-1}+1}^{r_n} \xi_i t_i = 1\Big\} = \Big( \sum_{r_{n-1}+1}^{r_n} \Big(\frac{t_i}{a_i^k}\Big)^2 \Big)^{-1},$$

where we must make a separate calculation for the last equality. That is, if $\lambda_i$, $\alpha_i$, $i=1,\dots,n$ are numbers we must show that

$$\min\Big\{\sum_i \lambda_i^2 : \sum_i \alpha_i \lambda_i = 1\Big\} = \Big(\sum_i \alpha_i^2\Big)^{-1} .$$

One inequality follows from the fact that for any such $(\lambda_i)$,

$$1 = (\sum_i \alpha_i \lambda_i)^2 \leq \sum_i \alpha_i^2 \sum_i \lambda_i^2$$

and the other by taking $\lambda_i = \alpha_i (\sum_j \alpha_j^2)^{-1}$.

But now, given $k \exists j \ni \sum_n (\frac{a_n^k}{a_n^j})^2 = M < \infty$ so

$$(b_n^k)^2 \sum_{i=r_{n-1}+1}^{r_n} (\frac{t_i}{a_i^j})^2 = (b_n^k)^2 \sum_{i=r_{n-1}+1}^{r_n} (\frac{a_i^k}{a_i^j})^2 (\frac{t_i}{a_i^k}) \leq M$$

and since, by (1) $Tx_{p_n^k} = t_{p_n^k} e_n$,

$$||e_n||_k \leq \frac{a^k_{p_n^k}}{|t_{p_n^k}|} = b_n^k \leq \sqrt{M} ||e_n||_k$$

so $K(b) = K((||e_n||_k))$. But E is nuclear so $\tilde{K}$ is nuclear and $(e_n)$ is a basis for $\tilde{K}$ so $\tilde{K} = K((||e_n||_k))$.

Now, for the first statement, let $G = \{x = \sum_i \xi_i x_i \in E: \sum_{i=r_{n-1}+1}^{r_n} \xi_i t_i = 0 \ \forall\, n \in \mathbb{N}\}$. Then G is closed in E so we have a quotient map $\pi: E \to E/G$. Set

$$y_n = \pi(\frac{x_{i(n)}}{t_{i(n)}}) \text{ so } \pi(x_i) = t_i y_n \text{ for } i \in (r_{n-1}, r_n]$$

which is (1). Moreover, if $(u_i)$ is the sequence biorthogonal to $(x_i)$, then $v_n = \sum_{i=p_{n-1}+1}^{p_n} t_i u_i$ defines a sequence biorthogonal to

$(y_n)$. Hence by the first part of the proof, $E/G \cong K(b)$ so we have the quotient map and (1) holds. ∎

(3.3.2) The argument for (3.3.1) seems much harder than the corresponding proof for block basic sequences and it may be that (3.3.1) has an easier proof.

(3.3.3) Theorem

If E,F are two nuclear Fréchet spaces neither of which is isomorphic to $\omega$, then there is a nuclear Fréchet space G with a continuous norm which is isomorphic to a quotient of E and a quotient of F.

Proof

By (3.1.3) and (3.1.4) we may assume that E,F have bases $(x_n)$, $(y_n)$ respectively with strict type $(d_0)$ representations a,b respectively, and $\lim_n \frac{a_n^k}{a_n^{k+1}} = \lim_n \frac{b_n^k}{b_n^{k+1}} = 0$.

Decompose $(x_n)$, $(y_n)$ into $(x_{m,\nu})$, $(y_{m,\nu})$, $m,\nu \in \mathbb{N}$ so that for each m, $(a^k_{m,\nu})_{k,\nu}$ and $(b^k_{m,\nu})_{k,\nu}$ are still of strict type $(d_0)$.

For each m we will select $p^1 > \dots > p^m$ and then $t_1,\dots,t_{p^m}$ according to (2.3.3) relative to the matrix $(a^k_{m,\nu})_{k,\nu}$ and define $c = (c^k_n)$ by

$$c^k_m = \frac{a^k_{m,p^k}}{|t_{p^k}|} > 0 \text{ (some of the dependence on m is not indicated)} .$$

We will also select $\tilde{p}^1 > \dots > \tilde{p}^m$ and then $\tilde{t}_1,\dots,\tilde{t}_{p^m}$ according to (2.3.3) relative to the matrix $(b^k_{m,\nu})_{k,\nu}$ and define $\tilde{c} = (\tilde{c}^k_n)$ by

$$\tilde{c}_m^k = \frac{b^k_{m,\tilde{p}^k}}{|\tilde{t}_{\tilde{p}^k}|} > 0 \text{ (some of the dependence on m is not indicated).}$$

Then by (2.3.3) and (3.3.1), $K(c)$, $K(\tilde{c})$ are quotients of E,F respectively.

It remains to select $p^k, \tilde{p}^k$. From the inequality of (2.3.3) we have

$$c_m^k \frac{a^{k+1}_{m,p^{k+1}}}{a^k_{m,p^k}} \leq c_m^{k+1} < c_m^k \frac{a^{k+1}_{m,p^k}}{a^k_{m,p^k}}$$

and

$$\tilde{c}_m^k \frac{b^{k+1}_{m,p^{k+1}}}{b^k_{m,p^{k+1}}} \leq \tilde{c}_m^{k+1} < \tilde{c}_m^k \frac{b^{k+1}_{m,p^k}}{b^k_{m,p^k}}$$

for $k = 1,\ldots,m$ and we can have $c_m^{k+1}$, $\tilde{c}_m^{k+1}$ as close to the right hand side of the inequality as we like.

Thus we may select $p^m$, $\tilde{p}^m$, $p^{m-1},\ldots, p^1$, $\tilde{p}^1$ so that $c_m^k \leq \tilde{c}_m^k \leq c_m^{k+1}$ for $m \geq k$. Hence $K(c) = K(\tilde{c})$. ∎

(3.4) Proposition

Every finite type power series space has a subspace isomorphic to some $L_f(\beta,\infty)$ space.

Proof

By passing to a subsequence if necessary we may consider a space $\Lambda_1(\alpha)$ where $\alpha$ satisfies

$$\frac{\alpha_{n+1}}{\alpha_{n+2}} \leq \frac{\alpha_n}{\alpha_{n+1}} \leq \frac{1}{n(n+1)} .$$

Let $r_n = \frac{n(n-1)}{2}$ and we will define a block basic sequence $(y_n)$ in $\Lambda_1(\alpha)$ by

$$y_n = \sum_{k=1}^{n} t_k^n e_{r_{n-1}+k}, \quad t_k^n = e^{\frac{1}{k}\alpha_{r_{n-1}+k}}, \quad k=1,\ldots,n .$$

Let $a_n^k = e^{-\frac{1}{k+1}\alpha_n}$ and we have, using our initial assumption,

$$\frac{1}{k+2}(\alpha_{r_{n-1}+k+1} - \alpha_{r_{n-1}+k}) \leq \frac{1}{k+1}\alpha_{r_{n-1}+k+1} - \frac{1}{k}\alpha_{r_{n-1}+k} <$$

$$< \frac{1}{k+1}(\alpha_{r_{n-1}+k+1} - \alpha_{r_{n-1}+k})$$

which immediately implies the condition of (2.2.3) so we have

$$q^k(y_n) = r_{n-1}+k \text{ for } k = 1,\ldots,n$$

so if we define $b = (b_n^k)$ by

$$b_n^k = e^{\frac{1}{k(k+1)}\alpha_{r_{n-1}+k}} \quad \forall n,k$$

then b is a representation of $(y_n)$. Also, from our initial assumption, if $c = (c_n^k)$ is defined by

$$c_n^k = e^{\alpha_{r_{n-1}+k}} \quad \forall n,k$$

then c is also a representation of $(y_n)$.

Finally, if we define $f: \mathbb{R} \to \mathbb{R}$ by

$$f(x) = \begin{cases} \alpha_n \left(\frac{x}{e^n}\right)^{\log\frac{\alpha_{n+1}}{\alpha_n}} & \text{for } n \leq \log x \leq n+1 \quad n = 1,2,\ldots \\ f(-x) & \text{for } x < 0 \end{cases}$$

then one can check that f is a rapidly increasing, odd, logarithmically convex function and $K(c) = L_f(e^{r_{n-1}}, \infty)$. ∎

## 4. Notes and Remarks

Conditions $(d_0)$, $(d_1)$ and $(d_2)$ (in slightly different form) were introduced by Dragilev [19]. Conditions $(d_3)$ and $(d_5)$ appear first in [29] and condition $(d_4)$ was introduced by the author and Robinson [34]. Condition $(d_6)$ appears here for the first time. Pseudo-regular bases were introduced by the author along with Crone and Robinson [12].

Proposition (1.4.3) was first proved in [31] and proposition (1.4.6) is due to Aytuna and Terzioglu [3] who use both results in connection with some questions regarding spaces of analytic functions.

Proposition (2.2.3) is the basic tool in our constructions of subspaces and has appeared in several papers in various forms. It was first proved in [28]. The analogue for quotients, proposition (2.3.3), is due to the author and Robinson [34].

The results in section 3 on subspaces, that is (3.1.1), (3.1.2), theorem 3.2 and (3.4) all appear in [28]. The choice of f in the proof of (3.4) greatly simplified the original argument. It was suggested by Zahariuta. Proposition (3.3.1) is due to the author and Robinson [34]. Proposition (3.1.3) and theorem (3.3.3) appear here for the first time.

# CHAPTER III

## SUBSPACES

### 1. The Fundamental Inequality

(1.1) Our main interest is to determine conditions which characterize those Köthe spaces which are isomorphic to a subspace of a given Köthe space. Always there are two conditions. The first is one of the type $(d_i)$ which determines a class of spaces and the second (which is I (6.2.4)) is a nuclearity condition which (by I (6.2.4) is clearly necessary.

The tools developed in Chapter II are for constructions so they represent sufficient conditions. For the necessity we have of course I (6.2.4) and an inequality which we now derive.

(1.2) We will consider a fixed nuclear Fréchet space E with basis $(x_n)$. If a is a strict $(d_0)$ representation of $(x_n)$ and

$y = \sum_i t_i x_i \in E$ we will define

$$q^k(y) = q^k(y;a) = \max\{q: \sup_i |t_i| a_i^k = |t_q| a_q^k\} .$$

Note that by the nuclearity, this set is non-empty and finite. Also, we observe that this is consistent with the definition in II (2.2.1), except that now we permit $\{t_1,\ldots,t_r\}$ to be infinite. It is clear that II (2.2.2) remains true.

(1.3) Proposition

In the context of (1.2) let F be a subspace of E with basis $(y_n)$ and set $q_n^k = q^k(y_n)$. Then $(y_n)$ has a representation b with

$b_n^k = |t_{q_n^k}| a_{q_n^k}^k$ and

$$\frac{a^j_{q^k_n}}{a^k_{q^k_n}} \leq \frac{b^j_n}{b^k_n} \leq \frac{a^j_{q^j_n}}{a^k_{q^j_n}} \quad \forall\ n,k,j \quad .$$

Proof

Let $y_n = \sum_i t^n_i x_i$ and set $b^k_n = |t^n_{q^k_n}| a^k_{q^k_n}$. By the nuclearity, $b = (b^k_n)$ is a representation of $(y_n)$. Thus we have

$$\frac{b^j_n}{b^k_n} = \frac{|t^n_{q^j_n}| a^j_{q^j_n}}{|t^n_{q^k_n}| a^k_{q^k_n}} \quad \forall\ n,k,j \ .$$

Now, by the definition of $q^k_n$, this quantity becomes larger if $q^k_n$ is replaced by $q^j_n$ and it becomes smaller if $q^j_n$ is replaced by $q^k_n$. The inequality follows immediately. ∎

2. Stable Power Series Spaces

(2.1) In this section we assume that $\alpha$ is a nuclear exponent sequence which satisfies

$$\sup_n \frac{\alpha_{2n}}{\alpha_n} = A < \infty \ .$$

It is easy to see, using the quasi-equivalence property, that this is equivalent to the condition that $\Lambda_\tau(\alpha) \cong \Lambda_\tau(\alpha) \times \Lambda_\tau(\alpha)$, $\tau = 1,\infty$. In the case of $L_f(\alpha,r)$ spaces, considered below, the condition can be different. This property is referred to as stability.

(2.2) Characterizations

(2.2.1) Theorem

Let F be a nuclear Fréchet space with basis $(y_n)$. Then F

is isomorphic to a subspace of $\Lambda_\infty(\alpha)$ iff

i) The basis $(y_n)$ is of type $(d_3)$

ii) $\forall$ nbd of 0 U in F and $\rho > 0$ $\exists$ nbd of 0 V in F $\ni$

$$d_n(V,U) \leq e^{-\rho\alpha_n} \text{ for n sufficiently large } .$$

Proof

First we show that the two conditions are necessary, so we assume that F is a subspace of $\Lambda_\infty(\alpha)$. The computation of Kolmogorov diameters in $\Lambda_\infty(\alpha)$ is done with I (6.3.2) and it is easy to check that ii) holds for $\Lambda_\infty(\alpha)$. By I (6.2.4) it holds for F.

For i) we apply (1.3) with $a_n^k = e^{k\alpha_n}$. This yields a representation b for $(y_n)$ $\ni$

$$\frac{b_n^{k+1}}{b_n^k} \leq e^{\alpha q_n^{k+1}} \leq \frac{b_n^{k+2}}{b_n^{k+1}} \quad \forall n,k$$

which is the $(d_3)$ condition for b.

Now we assume that i), ii) hold and we will construct a basic sequence $(z_m)$ in $\Lambda_\infty(\alpha)$ and a matrix b which is a representation for $(z_m)$ and a permutation of $(y_n)$. This will complete the proof.

Let c be a $(d_3)$ representation of $(y_n)$. Then by ii) and the basic properties of diameters we have a permutation $\pi$ of $\mathbb{N}$ and $j \in \mathbb{N}$ $\ni$

$$\frac{c_{\pi(n)}^1}{c_{\pi(n)}^{j+1}} \leq e^{-A\alpha_n} \quad \text{for n sufficiently large} \quad .$$

We can easily adjust c so that this holds $\forall n$ and c is still a $(d_3)$ representation for $(y_n)$. Moreover, this continues to be the case if we replace c by $(c_n^{(k-1)j+1})_{n,k}$ and then apply II (1.5.1). Thus we obtain a representation b of a permutation of $(y_n) \ni$

$$\frac{b_n^1}{b_n^2} \le e^{-A\alpha_n} \le e^{-\alpha_{2n-1}} \quad \forall n \tag{1}$$

and

$$\left(\frac{b_n^{k+1}}{b_n^k}\right)^A \le \frac{b_n^{k+2}}{b_n^{k+1}} \quad \forall\ n,k \quad . \tag{2}$$

Now, the coordinate basis $(e_n)$ in $\Lambda_\infty(\alpha)$ has the strict $(d_0)$ representation $(e^{k\alpha_n})$. We will fix $m \in \mathbb{N}$ and generally suppress it in our notation when the meaning is clear.

Let $x_\nu = e_{2^{\nu-1}(2m-1)}$, $\beta_\nu = \alpha_{2^{\nu-1}(2m-1)}$, $a_\nu^k = e^{k\beta_\nu}$, $k,\nu \in \mathbb{N}$ so that $(x_\nu)$ is a subsequence of $(e_n)$ (with no repetitions as m varies) and $(a_\nu^k)$ is a strict type $(d_0)$ representation of $(x_\nu)$. We will define

$$z_m = \sum_{k=1}^{m} t_{q^k} x_{q^k}$$

so we must select $q^k$, $t_{q^k}$.

First we show that we can select $q^1,\ldots,q^m \ni$

$$e^{\beta_{q^k}} = \frac{a_{q^k}^{k+1}}{a_{q^k}^k} \le \frac{b_m^{k+1}}{b_m^k} < \frac{a_{q^{k+1}}^{k+1}}{a_{q^{k+1}}^k} = e^{\beta_{q^{k+1}}}, \quad k = 1,\ldots,m-1 \ . \tag{3}$$

By (1) we may take $q^1 = 1$ and the left inequality holds. Suppose that we have chosen $q^1,\ldots,q^k \ni$ the left inequality and also the right inequality holds. Since $\lim_\nu \beta_\nu = \infty$ we can let $q^{k+1}$ be the smallest index $\ni$ the right inequality holds. Then $q^k < q^{k+1}$ and by (2),

$$e^{\beta_{q^{k+1}}} \leq e^{A\beta_{q^{k+1}-1}} \leq \left(\frac{b_m^{k+1}}{b_m^k}\right)^A \leq \frac{b_m^{k+2}}{b_m^{k+1}}$$

so the construction of $q^1,\ldots,q^m$ is complete.

Now set

$$t_{q^k} = \frac{b_m^k}{a_{q^k}^k} \qquad k = 1,\ldots,m$$

and it follows from (3) that

$$\frac{a_{q^k}^{k+1}}{a_{q^{k+1}}^{k+1}} \leq \frac{t_{q^{k+1}}}{t_{q^k}} < \frac{a_{q^k}^k}{a_{q^{k+1}}^k}, \qquad k = 1,\ldots,m-1 \ .$$

But this is exactly the condition of II (2.2.3) (with $j_\ell = \ell$) so if $||\cdot||_k$ is the "sup. norm" in $\Lambda_\infty(\alpha)$ it follows from II (2.2.3) that

$$||z_m||_k = |t_{q^k}| a_{q^k}^k = b_m^k, \qquad k = 1,\ldots,m \ .$$

Clearly, $(z_m)$ is a basic sequence in $\Lambda_\infty(\alpha)$ and b is then a representation of it. ∎

(2.2.2) Theorem

Let F be a nuclear Fréchet space with basis $(y_n)$. Then F is isomorphic to a subspace of $\Lambda_1(\alpha)$ iff

i) The basis $(y_n)$ is type $(d_5)$

ii) $\forall$ nbd of 0 U in F $\exists$ nbd of 0 V in F and $\rho > 0 \ni$

$$d_n(V,U) \le e^{-\rho\alpha_n} \text{ for } n \text{ sufficiently large .}$$

Proof

First we show that the two conditions are necessary so we assume that F is a subspace of $\Lambda_1(\alpha)$. The computation of Kolmogorov diameters in $\Lambda_1(\alpha)$ is done by I (6.3.2) and it is easy to check that ii) holds for $\Lambda_1(\alpha)$. By I (6.2.4) it holds for F.

For i) we apply (1.3) with $a_n^k = e^{-\frac{\alpha_n}{k}}$. This yields a representation b for $(y_n)$ ∍

$$\log\frac{b_n^{k+1}}{b_n^k} \le (\frac{1}{k} - \frac{1}{k+1})\alpha_{q_n^{k+1}} = \frac{k+2}{k}(\frac{1}{k+1} - \frac{1}{k+2})\alpha_{q_n^{k+1}} \le 3\log\frac{b_n^{k+2}}{b_n^{k+1}}$$

which is the $(d_5)$ condition for b with M = 3.

Now we assume that i) and ii) hold and we will construct a basic sequence $(z_m)$ in $\Lambda_1(\alpha)$ and a matrix b which is a representation for $(z_m)$ and a permutation of $(y_n)$. This will complete the proof.

Let c be a $(d_5)$ representation of $(y_n)$ so we have $M \ge 1$. By ii) and the basic properties of diameters we have a permutation $\pi$ of $\mathbb{N}$, $j \in \mathbb{N}$ and $\rho > 0$ ∍

$$\frac{c^1_{\pi(n)}}{c^{j+1}_{\pi(n)}} \le e^{-\rho\alpha_n} \quad \text{for } n \text{ sufficiently large .}$$

By decreasing $\rho$ if necessary we can be sure that this holds for all n. Now if we replace c by $(c_n^{(k-1)j+1})_{n,k}$ then the $(d_5)$ condition holds, but with M replaced by $M_1 = M^j$. Thus we obtain a representation b of a permutation of $(y_n)$ ∍

$$\frac{b_n^1}{b_n^2} \le e^{-\rho\alpha_n} \le e^{-\frac{\rho}{A}\alpha_{2n-1}} \quad \forall n \tag{4}$$

and

$$\frac{b_n^{k+1}}{b_n^k} \le \left(\frac{b_n^{k+2}}{b_n^{k+1}}\right)^{M_1} \quad \forall n,k \tag{5}$$

Now let $(\rho_k)$ be a strictly increasing unbounded sequence of positive numbers $\ni$

$$\frac{1}{\rho_1} \le \frac{\rho}{A} \quad \text{and} \quad \frac{\frac{1}{\rho_{k+1}} - \frac{1}{\rho_{k+2}}}{\frac{1}{\rho_k} - \frac{1}{\rho_{k+1}}} \le \frac{1}{AM_1} \ . \tag{6}$$

To see that the second relation is possible, ignore $\rho_{k+2}$ and, given $\rho_k$ take $\rho_{k+1}$ sufficiently large.

Now, the coordinate basis $(e_n)$ in $\Lambda_1(\alpha)$ has the strict $(d_0)$ representation $(e^{-\frac{1}{\rho_k}\alpha_n})$. We will fix m and generally suppress it in our notation when the meaning is clear.

Let $x_\nu = e_{2^{\nu-1}(2m-1)}$, $\beta_\nu = \alpha_{2^{\nu-1}(2m-1)}$, $a_\nu^k = e^{-\frac{1}{\rho_k}\beta_\nu}$,

$k,\nu \in \mathbb{N}$ so that $(x_\nu)$ is a subsequence of $(e_n)$ (with no repetitions as m varies) and $a_\nu^k$ is a strict type $(d_0)$ representation of $(x_\nu)$. We will define

$$z_m = \sum_{k=1}^{m} t_{q^k} x_{q^k}$$

so we must select $q^k$, $t_{q^k}$ .

First we show that we can select $q^1,\dots,q^m \ni$

$$e^{\left(\frac{1}{\rho_k}-\frac{1}{\rho_{k+1}}\right)\beta_{q^k}} = \frac{a^{k+1}_{q^k}}{a^{k}_{q^k}} \le \frac{b^{k+1}_m}{b^k_m} < \frac{a^{k+1}_{q^{k+1}}}{a^{k}_{q^{k+1}}} = e^{\left(\frac{1}{\rho_k}-\frac{1}{\rho_{k+1}}\right)\beta_{q^{k+1}}}, \quad k=1,\dots,m-1 \tag{7}$$

By (4) and (6) we may take $q^1 = 1$ and the left inequality holds. Suppose that we have chosen $q^1,\dots,q^k \ni$ the left inequality and also the right inequality holds. Since $\rho_k < \rho_{k+1}$ and $\lim_\nu \beta_\nu = \infty$ we can let $q^{k+1}$ be the smallest index $\ni$ the right inequality holds. Then by (5) and (6),

$$\log\left(\frac{a^{k+2}_{q^{k+1}}}{a^{k+1}_{q^{k+1}}}\right) = \left(\frac{1}{\rho_{k+1}}-\frac{1}{\rho_{k+2}}\right)\beta_{q^{k+1}} \le \frac{1}{AM_1}\,\frac{\beta_{q^{k+1}}}{\beta_{q^{k+1}-1}}\left(\frac{1}{\rho_k}-\frac{1}{\rho_{k+1}}\right)\beta_{q^{k+1}-1}$$

$$\le \frac{1}{M_1}\log\frac{b^{k+1}_m}{b^k_m} \le \log\frac{b^{k+2}_m}{b^{k+1}_m}$$

so the construction of $q^1,\dots,q^m$ is complete.

Now set

$$t_{q^k} = \frac{b^k_m}{a^k_{q^k}} \qquad k = 1,\dots,m$$

and it follows from (7) that

$$\frac{a^{k+1}_{q^k}}{a^{k+1}_{q^{k+1}}} \le \frac{t_{q^{k+1}}}{t_{q^k}} < \frac{a^{k}_{q^k}}{a^{k}_{q^{k+1}}}, \quad k = 1,\dots,m-1 \ .$$

But this is exactly the condition of II (2.2.3) (with $j_\ell = \ell$) so if $||\cdot||_k$ is the sup. norm in $\Lambda_1(\alpha)$ it follows from II (2.2.3) that

$$||z_m||_k = |t_{q^k}| a^k_{q^k} = b^k_m , \quad k = 1,\dots,m .$$

Clearly $(z_m)$ is a basic sequence in $\Lambda_\infty(\alpha)$ and b is then a representation of it. ∎

(2.2.3) We remark that although the proofs of (2.2.1) and (2.2.2) are nearly identical there are important differences. First, the fact that condition ii) in (2.2.1) is so much stronger than condition ii) in (2.2.2) requires an additional argument. Also, II (1.5.1) which was used in (2.2.1) is not available with the $(d_5)$ condition. Both of these are taken care of by the argument leading to (6) which cannot be done for the infinite type power series space.

(2.3) Some Consequences

(2.3.1) Corollary

A nuclear Fréchet space F with basis is isomorphic to a subspace of (s) iff it has a basis of type $(d_3)$ iff every basis is type $(d_3)$.

<u>Proof</u>

$(s) = \Lambda_\infty(\alpha)$ with $\alpha_n = \log n$ in which case condition ii) of (2.2.1) is equivalent to nuclearity ([53]). ∎

(2.3.2) Corollary

If E is a nuclear Fréchet space with a basis of type $(d_3)$, then every basic sequence in E is type $(d_3)$.

Proof

By (2.3.1), E is a subspace of (s) so every basic sequence in E is a basic sequence in (s) so by (2.2.1), it is type $(d_3)$.

∎

(2.3.3) Corollary

If E is a nuclear Fréchet space with a basis of type $(d_3)$ then no basic sequence in E is type $(d_2)$.

Proof

If $(y_n)$ is a basic sequence in E then by (2.3.2) it is $(d_3)$ so by II (1.4.1) it is $(d_1)$ and it is easy to see from the definitions and the nuclearity that a basis cannot be both type $(d_1)$ and type $(d_2)$.

∎

(2.3.4) Corollary

If E is isomorphic to a subspace of $\Lambda_1(\alpha)$ and E has a basis of type $(d_2)$ then E is isomorphic to a finite type power series space.

Proof

Immediate from (2.2.2) and II (1.4.3).

∎

(2.3.5) Corollary

Let E be isomorphic to a subspace of $\Lambda_1(\alpha)$ and assume that E has a basis $(x_n)$. Then either $(x_n)$ is type $(d_1)$ or there is a subsequence $(j_n) \ni (x_{j_n})$ is a basis for a (complemented) subspace of E isomorphic to a finite type power series space.

Proof

Immediate from (2.2.2) and II (1.4.6)

∎

(2.4) Power Series Subspaces of Power Series Spaces

(2.4.1) In this section we continue to assume that $\alpha$ is a stable nuclear exponent sequence and we let $\beta$ be a nuclear exponent sequence not assumed to be stable. In both cases, the type is determined by the context.

(2.4.2) Corollary

Let $\tau = 1$ or $\infty$. Then $\Lambda_\tau(\beta)$ is isomorphic to a subspace of $\Lambda_\tau(\alpha)$ iff

$$\sup_n \frac{\alpha_n}{\beta_n} < \infty . \tag{8}$$

Proof

The coordinate basis is clearly type $(d_3)$ or $(d_5)$ as the case may be and condition ii) of (2.2.1) or (2.2.2) is in both cases equivalent to (8). ∎

(2.4.3) Corollary

The space $\Lambda_1(\beta)$ is not isomorphic to a subspace of $\Lambda_\infty(\alpha)$.

Proof

The coordinate basis in $\Lambda_1(\beta)$ is type $(d_2)$ (see II (1.2.2)) so by (2.3.3) it cannot be a basic sequence in $\Lambda_\infty(\alpha)$. ∎

(2.4.4) Corollary

The space $\Lambda_\infty(\beta)$ is isomorphic to a subspace of $\Lambda_1(\beta)$ iff (8) holds.

Proof

First let us apply condition ii) of (2.2.2). We can compute Kolmogorov diameters in $\Lambda_\infty(\beta)$ by I (6.3.2), so the condition is,

$$\forall\, k\ \exists\, j \text{ and } \rho > 0 \ni e^{(k-j)\beta_n} \le e^{-\rho\alpha_n} \text{ for n sufficiently large,}$$

which is clearly equivalent to (8). Thus it remains only to observe that a basis of type $(d_3)$ is clearly $(d_5)$ (take M = 1). ∎

(2.4.5) Lemma

Given $\beta$ ∃ (stable) $\alpha$ ∋ $\Lambda_\tau(\beta)$ is isomorphic to a subspace of $\Lambda_\tau(\alpha)$, $\tau = 1,\infty$.

Proof

If $\tau = \infty$ we can simply take $\alpha_n = \log n$ and apply, say, (2.3.1).

Let $\tau = 1$ and let $\gamma$ be any stable nuclear exponent sequence of finite type. We will consider Köthe spaces K(a) with $a = (a^k_{m,\nu})$, that is, the coordinate index n is replaced by the two-dimensional index $(m,\nu)$. Now define matrices a, b, c by

$$a^k_{m,\nu} = e^{-\frac{1}{k}(\beta_m+\gamma_\nu)}, \quad b^k_{m,\nu} = a^k_{m,2\nu}, \quad c^k_{m,\nu} = a^k_{m,2\nu-1} .$$

Clearly, $K(a) \cong K(b) \times K(c)$ and from the stability of $\gamma$, $K(b) \cong K(c) \cong K(a)$ so $K(a) \cong K(a) \times K(a)$. Next, the numbers $(\beta_m+\gamma_\nu)_{m,\nu}$ can be arranged into a sequence $\alpha = (\alpha_n)$ so $K(a) \cong \Lambda_1(\alpha)$ and the nuclearity follows from the fact that given k we can find j ∋

$$\sum_{m,\nu} \frac{a^k_{m,\nu}}{a^j_{m,\nu}} = \sum_{m,\nu} e^{(\frac{1}{j}-\frac{1}{k})(\beta_m+\gamma_\nu)} = \sum_m e^{(\frac{1}{j}-\frac{1}{k})\beta_m} \sum_\nu e^{(\frac{1}{j}-\frac{1}{k})\gamma_\nu} < \infty .$$

Hence (see 2.1), $\alpha$ is a stable nuclear exponent sequence. Finally, if $(m,\nu)$ is restricted to $(m,1)$ we obtain a space isomorphic to $\Lambda_1(\beta)$. ∎

(2.4.6) Corollary

The results of (2.3.4), (2.3.5) and (2.4.3) remain true if we drop the assumption that $\alpha$ is stable.

Proof

If E is isomorphic to a subspace of a power series space then by (2.4.5) it is isomorphic to a subspace of a stable power series space. ∎

(2.5) $L_f(\alpha,r)$ Subspaces of Power Series Spaces

(2.5.1) Corollary

If $r \leq 0$ then $L_f(\beta,r)$ is not isomorphic to a subspace of $\Lambda_\infty(\alpha)$.

Proof

If $r \leq 0$ then the coordinate basis in $L_f(\beta,r)$ is type $(d_2)$ so by (2.3.3) the result follows. ∎

(2.5.2) Proposition

If $0 < r \leq \infty$ then $L_f(\beta,r)$ is isomorphic to a subspace of $\Lambda_\infty(\alpha)$ iff

$$\exists \text{ positive } \rho_0 < r \ni \sup_n \frac{\alpha_n}{f(\rho_0\beta_n)} < \infty \quad . \tag{9}$$

Proof

In this case the coordinate basis in $L_f(\beta,r)$ is type $(d_1)$ and hence by II (1.4.1) it is type $(d_3)$. Thus we need only show that condition ii) of (2.2.1) is equivalent to (9). Computing the Kolmogorov diameters in $L_f(\beta,r)$ by I (6.3.2) we conclude that condi tion ii) of (2.2.1) is equivalent to,

$$\forall \text{ positive } \rho_1 < r \text{ and } \rho > 0 \ \exists \text{ positive } \rho_2 < r \ni$$

$$\rho\alpha_n \leq f(\rho_2\beta_n) - f(\rho_1\beta_n) \text{ for n sufficiently large.}$$

This clearly implies (9) whereas if (9) holds, given $\rho,\rho_1$ we choose $\rho_2 \ni \max(\rho_0,\rho_1) < \rho_2 < r$ and using the fact (twice) that f is rapidly increasing, we have $C > 0$ from (9) $\ni$ for n sufficiently large,

$$2^{\rho\alpha_n} \leq Cf(\rho_0\beta_n) \leq f(\rho_2\beta_n) \leq 2f(\rho_2\beta_n) - 2f(\rho_1\beta_n)$$

which gives our result. ∎

(2.5.3) Proposition

If $0 < r \leq \infty$ then $L_f(\beta,r)$ is isomorphic to a subspace of $\Lambda_1(\alpha)$ iff (9) holds.

Proof

As in (2.5.2) the coordinate basis in $L_f(\beta,r)$ is $(d_3)$ and so it is type $(d_5)$. Thus we need only show that (9) is equivalent to condition ii) of (2.2.2) which we see, by computing the Kolmogorov diameters, is equivalent to,

$$\forall \text{ positive } \rho_1 < r \; \exists \text{ positive } \rho_2 < r \text{ and } \rho > 0 \ni$$

$$\rho\alpha_n \leq f(\rho_2\beta_n) - f(\rho_1\beta_n) \text{ for n sufficiently large}$$

This clearly implies (9) whereas if (9) holds, the condition we derived in (2.5.2) is actually stronger than this one. ∎

(2.5.3) Proposition

If $-\infty < r \leq 0$ then $L_f(\beta,r)$ is not isomorphic to a subspace of $\Lambda_1(\alpha)$.

Proof

Let b be the representation for the coordinate basis in $L_f(\beta,r)$ given by

$$b_n^k = e^{f(\rho_k\beta_n)}, \quad \rho_k \nearrow r$$

Consider any $\rho_1 < \rho_2 < \rho_3 < r$. Then since f is strongly increasing,

$$\sup_n \frac{f(\rho_2\beta_n)-f(\rho_1\beta_n)}{f(\rho_3\beta_n)-f(\rho_2\beta_n)} \geq \sup_n \frac{f(\rho_2\beta_n)-f(\rho_1\beta_n)}{-f(\rho_2\beta_n)} = \sup_n \frac{f(\rho_1\beta_n)}{f(\rho_2\beta_n)} -1 = \infty$$

so by II (1.3.3) the coordinate basis in $L_f(\beta,r)$ is not $(d_5)$ so by (2.2.2) $L_f(\beta,r)$ is not isomorphic to a subspace of $\Lambda_1(\alpha)$.

∎

3. Unstable Power Series Spaces

(3.1) In this section we assume that $\lim_n \frac{\alpha_{n+1}}{\alpha_n} = \infty$ and say that $\alpha$ is unstable. We have no complete characterizations but we do have some particular results based on methods which should lead to more general results. This is an area in which there is room for research.

(3.2) Lemma

Let $\sigma,\tau = 1$ or $\infty$ and suppose that $\Lambda_\sigma(\beta)$ is isomorphic to a subspace of $\Lambda_\tau(\alpha)$. Then $\exists$ a Köthe space K(b) and an increasing unbounded sequence of positive integers $(\nu_n) \ni \Lambda_\sigma(\beta) \cong K(b)$ and

$$\frac{b_n^j}{b_n^j} = e^{\rho_{kj}\alpha_{\nu_n}} \quad \forall\, n,k,j$$

where $\rho_{kj} = \frac{1}{k} - \frac{1}{j}$ or j-k according as $\tau = 1$ or $\infty$.

Proof

Let $a_n^k = e^{-\frac{1}{k}\alpha_n}$ or $e^{k\alpha_n}$ according as $\tau = 1$ or $\infty$ so a is a representation of the coordinate basis in $\Lambda_\tau(\alpha)$. Let $(y_n)$ be the coordinate basis in $\Lambda_\sigma(\beta)$ and apply (1.3) to obtain $b, (q_n^k)$ as a representation of $(y_n)$ with $b_n^k = |t^n_{q_n^k}| a^k_{q_n^k}$. Let $c_n^k =$ $e^{-\frac{1}{k}\beta_n}$ or $e^{k\beta_n}$ according as $\sigma = 1$ or $\infty$ so c is also a representation of $(y_n)$ so we have subsequences of indices $(j_k)$, $(\ell_k) \ni \forall k$,

$$b_n^{j_k} \leq c_n^{\ell_k} \leq b_n^{j_{k+1}}, \text{ n sufficiently large .}$$

Putting this together with (1.3) yields, $\forall k$,

$$\frac{c_n^{\ell_{k-1}}}{c_n^{\ell_{k-2}}} \leq \frac{b_n^{j_k}}{b_n^{j_{k-2}}} \leq \frac{a^{j_k}_{q_n^{j_k}}}{a^{j_{k-2}}_{q_n^{j_k}}} \text{ and } \frac{a^{j_{k+2}}_{q_n^{j_{k+1}}}}{a^{j_{k+1}}_{q_n^{j_{k+1}}}} \leq \frac{b_n^{j_{k+2}}}{b_n^{j_{k+1}}} - \frac{c_n^{\ell_{k+2}}}{c_n^{\ell_k}} ,$$

n sufficiently large .

This implies that we have positive constants $L_k$, $M_k \ni$

$$L_k\beta_n \leq \alpha_{q_n^{j_k}} \text{ and } M_k\alpha_{q_n^{j_{k+1}}} \leq \beta_n \quad \forall n$$

and since $q_n^k \leq q_n^{k+1}$ (see (1.2)) it follows that

$$\sup_n \frac{\alpha_{q_n^{k+1}}}{\alpha_{q_n^k}} < \infty \quad \forall k$$

and so by the assumption of unstability, $\forall k \; \exists n_k \ni q_n^{k+1} = q_n^k$ for

$n \geq n_k$. Thus given j,k we can find $n_{kj} \ni q_n^\ell = q_n^{\ell+1}$ $\forall$ $\ell < \max(k,j)$, $n \geq n_{kj}$. Hence, by the definition of b,

$$\frac{b_n^j}{b_n^k} = \frac{\left|t^n_{q_n^j}\right| \left|a^j_{q_n^j}\right|}{\left|t^n_{q_n^k}\right| \left|a^k_{q_n^k}\right|} = \frac{a^j_{q_n^1}}{a^k_{q_n^1}} = e^{\rho_{jk}\alpha_{q_n^1}} \;, \quad n \text{ sufficiently large}$$

Clearly we can adjust b so that this holds for all n,k,j and b is still a representation of $(y_n)$. Finally, by the nuclearity $\lim_n q_n^1 = \infty$ so we can rearrange it to an increasing unbounded sequence of positive integers $(\nu_n)$. We must apply this same rearrangement to b.

∎

(3.3) Corollary

There is no infinite type power series space isomorphic to a subspace of $\Lambda_1(\alpha)$.

Proof

If we had such a $\Lambda_\infty(\beta)$ then applying (3.2) with $j = 2k$ yields that b is type $(d_2)$ which is a contradiction since it must be $(d_1)$ and it cannot be both.

∎

(3.4) Lemma

Let $(y_n)$ be a basic sequence in a Fréchet space E and let $(z_n)$ be a sequence in E, $(||\cdot||_k)$ a fundamental system of norms $\ni$

$$\sum_n \frac{||y_n - z_n||_j}{||y_n||_k} < \infty \quad \forall j,k \quad .$$

Then $\exists\, n_0 \ni (z_n)_{n \geq n_0}$ is a basic sequence in E and the map $y_n \rightsquigarrow z_n$, $n \geq n_0$ extends to an isomorphism of the subspaces generated by $(y_n)_{n \geq n_0}$, $(z_n)_{n \geq n_0}$.

Proof

It is well known ([8]) that the space generated by $(y_n)$ has a fundamental system of norms $(|\cdot|_k) \ni (y_n)$ is a basis wrt each $|\cdot|_k$. We then extend $(|\cdot|_k)$ to a fundamental system of norms for E so we have subsequences of indices $(j_k)$ and $(\ell_k)$ and positive numbers $A_k$, $B_k \ni$

$$|x|_{j_k} \leq A_k||x||_{\ell_k} \leq B_k||x||_{j_{k+1}} \quad \forall\ k \text{ and } x \in E\ .$$

Then, for each k, by our assumption,

$$\sum_n \frac{|y_n-z_n|_{j_k}}{|y_n|_{j_k}} \leq \frac{A_{k-1}A_k}{B_{k-1}} \sum_n \frac{||y_n-z_n||_{\ell_k}}{||y_n||_{\ell_{k-1}}} < \infty$$

so by standard stability results our conclusion follows for each $|\cdot|_k$ and hence since each $|\cdot|_k$ is a norm, it holds for E (see [8]).

∎

(3.5) Lemma

In the context of (3.2), it follows moreover that

$\exists n_0 \ni \nu_n < \nu_{n+1}\ \forall_{n \geq n_0}$.

Proof

Our first estimate is that $n \leq \nu_n$. Indeed, it follows from I(6.2.2), (6.3.2) and the conclusion of (3.2) that $\sup_n \frac{\alpha_n}{\alpha_{\nu_n}} < \infty$

so by the instability, $\alpha_n \leq \alpha_{\nu_n}$ for n sufficiently large and $\alpha$ is strictly increasing so $n \leq \nu_n$ for n sufficiently large.

Now we have strictly increasing sequences $(r_n)$ and $(\mu_n)$ of integers with $0 = r_0$ and

$$\nu_m = \mu_n \text{ for } r_{n-1} < m \le r_n, \quad n = 1,2,\ldots$$

Let q, j, k, ℓ be indices, $r_{n-1} < m \le r_n$. Then,

$$\frac{|t_q^m|a_q^j}{|t^m_{q^k_m}|a^k_{q^k_m}} = \frac{|t_q^m|a_q^\ell}{|t^m_{q^k_m}|a^k_{q^k_m}}\frac{a_q^j}{a_q^\ell} \le \frac{|t^m_{q^\ell_m}|a^\ell_{q^\ell_m}}{|t^m_{q^k_m}|a^k_{q^k_m}}\frac{a_q^j}{a_q^\ell} = \frac{b_m^\ell}{b_m^k}\frac{a_q^j}{a_q^\ell} =$$

$$= e^{\rho_{k\ell}\alpha_{\nu_m}+\rho_{\ell j}\alpha_q} = e^{\rho_{k\ell}\alpha_{\mu_n} + \rho_{\ell j}\alpha_q}$$

Let $0 < \rho < \min(|\rho_{k,k-1}|,|\rho_{j+1,\ell}|)$, and we assume $q \ne \mu_n$. First, if $q < \mu_n$ take $\ell = k-1$ and we have $\rho_{k,k-1} < 0$ so by the instability,

$$\frac{|t_q^m|a_q^j}{|t^m_{q^k_m}|a^k_{q^k_m}} \le e^{\rho_{k,k-1}\alpha_{\mu_n}+\rho_{k-1,j}\alpha_q} < e^{-\frac{1}{2}\rho\alpha_{\mu_n}} \text{ for n sufficiently large}$$

and if $q > \mu_n$ take $\ell = j + 1$ and $\rho_{j+1,j} < 0$ so by the instability,

$$\frac{|t_q^m|a_q^j}{|t^m_{q^k_m}|a^k_{q^k_m}} \le e^{\rho_{k,j+1}\alpha_{\mu_n}+\rho_{j+1,j}\alpha_q} < e^{-\frac{1}{2}\rho\alpha_{\mu_n}} \text{ for n sufficiently large.}$$

Hence if we take $z_m = |t^m_{q^1_m}|e_{q^1_m}$, then recalling that, given k, j, $q^k_m = q^1_m$ for m sufficiently large, we have $\rho > 0$ depending only on k, j ∍

$$\sum_n \frac{||y_n-z_n||_j}{||y_n||_k} = \sum_n \sum_{m=r_{n-1}+1}^{r_n} \frac{||y_m-z_m||_j}{||y_m||_k} = \sum_n \sum_{m=r_{n-1}+1}^{r_n} \frac{\sup_{q\ne q^1_m} |t_q^m|a_q^j}{|t^m_{q^k_m}|a^k_{q^k_m}}$$

$$= \sum_n \sum_{m=r_{n-1}+1}^{r_n} \frac{\sup_{j,\, q \neq q_m^j} |t_q^m| a_q^j}{|t_{q_m^k}^m| a_{q_m^k}^k} \leq \sum_n (r_n - r_{n-1}) e^{-\frac{1}{2}\rho\alpha_{\mu_n}}$$

$$\leq \sum_n \nu_{r_n} e^{-\frac{1}{2}\rho\alpha_{\mu_n}} = \sum_n \mu_n e^{-\frac{1}{2}\rho\alpha_{\mu_n}} \leq \sum_n n e^{-\frac{1}{2}\rho n} < \infty \quad .$$

Hence by (3.4), $(e_{q_m^1}) = (e_{\nu_m})$ is a basic sequence after finitely many terms have been deleted so $\nu_m$ can have no repetitions. ∎

(3.6) Corollary

$\Lambda_1(\beta)$ is isomorphic to a subspace of $\Lambda_1(\alpha)$ iff $\exists$ a subsequence $(j_n)$ of indices $\ni$

$$0 < \inf_n \frac{\alpha_{j_n}}{\beta_n} \leq \sup_n \frac{\alpha_{j_n}}{\beta_n} < \infty \qquad (10)$$

<u>Proof</u>

One way is clear and the other is immediate from (3.5). ∎

(3.7) Corollary

$\Lambda_\infty(\beta)$ is isomorphic to a subspace of $\Lambda_\infty(\alpha)$ iff $\exists$ a subsequence $(j_n)$ of indices $\ni$ (10) holds.

<u>Proof</u>

One way is clear and the other is immediate from (3.5). ∎

## 4. Weakly Stable Power Series Spaces

(4.1) In this section we begin the study of the case

$\sup_n \frac{\alpha_{n+1}}{\alpha_n} = A < \infty$. From the quasi-equivalence property it is easy to see that this condition is equivalent to the corresponding power series space being isomorphic to its closed hyperplanes. We refer to this property as weak stability.

Here our results are really fragmentary. In the first place we deal only with subspaces of $\Lambda_1(\alpha)$. The main result, (4.2), gives only sufficient conditions which may or may not be necessary. The rest of the results are about very special situations, e.g. when $\alpha_n = 2^n$.

But we do get some non-trivial information such as (4.3). We have a beginning of a theory analogous to (2.4). Finally we give a construction of a basic sequence that cannot be done with a block basic sequence. This is the first such example and answers a question raised earlier [29].

It is clear that further research here should lead to several results but there are serious difficulties and the outcome cannot be predicted. What is needed are new methods. We give some now and hope that others are forthcoming.

(4.2) Proposition

Let F be a nuclear Fréchet space with a basis that has a representation b which is $(d_5)$ with $M < A$ and satisfies

$$\left(\frac{b_n^{k+1}}{b_n^k}\right)^A \leq \frac{b_{n+1}^{k+1}}{b_{n+1}^k} \quad \forall n,k . \tag{11}$$

Then F is isomorphic to a subspace of $\Lambda_1(\alpha)$.

Proof

Applying (11) and then ($d_5$) successively we obtain

$$\frac{b_n^k}{b_n^{k+1}} \le \left(\frac{b_1^1}{b_1^2}\right)^{\frac{A^{n-1}}{M^{k-1}}} \qquad \forall\, n,k \;.$$

Also, it follows from the weak stability that $\alpha_n \le A^{n-1}\alpha_1$ so

$$\exists\ \delta > 0 \ni e^{\delta\alpha_n} \le \frac{b_n^2}{b_n^1} \qquad \forall\, n \quad .$$

Moreover, if we define $\varepsilon_n \in (0,1)$ by $\varepsilon_n = \max\limits_{1\le k\le n} \left(\frac{b_n^k}{b_n^{k+1}}\right)^{\frac{1}{n+1}}$ then

$$\sum_n \varepsilon_n^{\frac{1}{M(M+1)}} < \infty \quad .$$

Now we choose an increasing unbounded sequence $(\rho_k) \ni$

$$\frac{1}{\rho_1} \le \delta\left(1-\frac{1}{M(M+1)}\right) \text{ and } \frac{1}{\rho_{k+1}} - \frac{1}{\rho_{k+2}} \le \frac{1}{A(M+1)}\left(\frac{1}{\rho_k} - \frac{1}{\rho_{k+1}}\right) \quad \forall k,$$

so if $a_n^k = e^{-\frac{1}{\rho_k}\alpha_n}$ then $a = (a_n^k)$ is a representation of $(e_n)$ in $\Lambda_1(\alpha)$. We will define a sequence $(y_n)$ in $\Lambda_1(\alpha)$ of the form

$$y_n = \sum_{k=1}^{n} t^n_{q^k_m}\, e_{q^k_n} \quad .$$

First we will define $q_n^k \ni q_n^1 = n$ and

$$\frac{a^{k+1}_{q^k_m}}{a^k_{q^k_n}} \le (\varepsilon_n)^{\frac{n+1}{M(M+1)}} \frac{b_n^{k+1}}{b_n^k} < \frac{b_n^{k+1}}{b_n^k} \le \frac{a^{k+1}_{q_n^{k+1}}}{a^k_{q_n^{k+1}}}, \quad k \le n \ .$$

To verify the left inequality with $q_n^1 = n$ we compute,

$$\frac{a_n^2}{a_n^1} \le e^{\frac{1}{\rho_1}\alpha_n} \le \left(e^{\delta\alpha_n}\right)^{1-\frac{1}{M(M+1)}} \le \left(\frac{b_n^2}{b_n^1}\right)^{1-\frac{1}{M(M+1)}} = \left(\frac{b_n^1}{b_n^2}\right)^{\frac{1}{M(M+1)}} \frac{b_n^2}{b_n^1} \le$$

$$(\varepsilon_n)^{\frac{n+1}{M(M+1)}} \frac{b_n^2}{b_n^1} \ .$$

Now we argue as usual by supposing that $q_n^1, \ldots, q_n^k$ have been chosen and then select $q_n^{k+1}$ to be the smallest index $\ni$ the right inequality holds so $q_n^k < q_n^{k+1}$. Now to verify the left inequality we compute,

$$\frac{a^{k+2}_{q_n^{k+1}}}{a^{k+1}_{q_n^{k+1}}} = e^{\left(\frac{1}{\rho_{k+1}} - \frac{1}{\rho_{k+2}}\right)\alpha_{q_n^{k+1}}} \le e^{\frac{1}{M+1}\left(\frac{1}{\rho_k} - \frac{1}{\rho_{k+1}}\right)\alpha_{q_n^{k+1}-1}} = \left(\frac{a^{k+1}_{q_n^{k+1}-1}}{a^{k}_{q_n^{k+1}-1}}\right)^{\frac{1}{M+1}}$$

$$< \left(\frac{b_n^{k+1}}{b_n^k}\right)^{\frac{1}{M} - \frac{1}{M(M+1)}} \le (\varepsilon_n)^{\frac{n+1}{M(M+1)}} \frac{b_n^{k+2}}{b_n^{k+1}} \ .$$

Thus we have $q_n^1,\ldots,q_n^n$. Clearly $q_n^1 < q_{n+1}^1$ and by our construction and (11),

$$\frac{a^{k+1}_{q_n^{k+1}-1}}{a^k_{q_n^{k+1}-1}} < \frac{b_n^{k+1}}{b_n^k} \leq \left(\frac{b_{n+1}^{k+1}}{b_{n+1}^k}\right)^{\frac{1}{A}} \leq \left(\frac{a^{k+1}_{q_{n+1}^{k+1}}}{a^k_{q_{n+1}^{k+1}}}\right)^{\frac{1}{A}} \quad \text{for } k = 1,\ldots,n$$

so by the definition of $a_n^k$, $\alpha_{q_n^{k+1}-1} < \frac{1}{A}\alpha_{q_{n+1}^{k+1}} \leq \alpha_{q_{n+1}^{k+1}-1}$ so

$q_n^{k+1} < q_{n+1}^{k+1}$. Hence, $q_n^k < q_{n+1}^k$ and, clearly, $q_n^k \leq q_n^{k+1}$ $k \leq n$.

Now we set

$$t^n_{q_n^k} = \frac{b_n^k}{a^k_{q_n^k}} \rho_n^k \quad \text{where } \rho_n^k = (\varepsilon_n)^{\frac{k-1}{M(M+1)}},\ k \leq n .$$

We must verify the condition of II (2.2.3) with $j_\ell = \ell$ so we compute

$$\frac{a^{k+1}_{q_n^k}}{a^{k+1}_{q_n^{k+1}}} < \frac{a^{k+1}_{q_n^k}}{a^{k+1}_{q_n^{k+1}}}(\varepsilon_n)^{-\frac{n}{M(M+1)}} \leq \frac{|t^n_{q_n^{k+1}}|}{|t^n_{q_n^k}|} < \frac{a^k_{q_n^k}}{a^k_{q_n^{k+1}}}(\varepsilon_n)^{\frac{1}{M(M+1)}} < \frac{a^k_{q_n^k}}{a^k_{q_n^{k+1}}} .$$

Hence by II (2.2.3) if $(||\cdot||_k)$ is the sequence of sup norms in $\Lambda_1(\alpha)$, then $||y_n||_k = b_n^k\rho_n^k$, $k \leq n$. Then

$$||y_n||_k \leq b_n^k \leq (\varepsilon_n)^{n+1} b_n^{k+1} < (\varepsilon_n)^{\frac{1}{M(M+1)}} b_n^{k+1} = \rho_n^{k+1} b_n^{k+1} = ||y_n||_{k+1},$$

$$k \leq n$$

so $K((||y_n||_k)) = K(b)$ and it remains only to show that $(y_n)$ is a basic sequence in $\Lambda_1(\alpha)$.

Fix k and let $z_n = t^n_{q^k_n} e_{q^k_n}$, $n \geq k$. Since $q^k_n < q^k_{n+1}$ and $t_{q^k_n} \neq 0$ it follows that $(z_n)$ is a basic sequence in the Banach space $(\Lambda_1(\alpha), ||\cdot||_k)^{\hat{}}$. Then in the computation of $||y_n - z_n||_k$ it follows from II (2.2.2) that the sup occurs at either $q^{k-1}_n$ or $q^{k+1}_n$. Hence,

$$\sum_{n\geq k} \frac{||y_n - z_n||_k}{||y_n||_k} = \sum_{n\geq k} \frac{1}{b^k_n \rho^k_n} \max\{|t^n_{q^{k-1}_n}| a^k_{q^{k-1}_n}, |t^n_{q^{k+1}_n}| a^k_{q^{k+1}_n}\}$$

$$= \sum_{n\geq k} \max\left\{\frac{b^{k-1}_n \rho^{k-1}_n a^k_{q^{k-1}_n}}{b^k_n \rho^k_n a^{k-1}_{q^{k-1}_n}}, \frac{b^{k+1}_n \rho^{k+1}_n a^k_{q^{k+1}_n}}{b^k_n \rho^k_n a^{k+1}_{q^{k+1}_n}}\right\}$$

$$\leq \sum_n \max\{\varepsilon_n^{\frac{n}{M(M+1)}}, \varepsilon_n^{\frac{1}{M(M+1)}}\} < \infty .$$

Hence as in (3.4) $\exists\, n_k \geq k \ni (y_n)_{n\geq n_k}$ is a basic sequence in $(\Lambda_1(\alpha), ||\cdot||_k)^{\hat{}}$. Since $q^1_n = n$ it follows that no $y_n$ is in the closed subspace of $(\Lambda_1(\alpha), ||\cdot||_k)^{\hat{}}$ generated by $(y_m)_{m>n}$ it follows that $(y_n)_{n\geq 1}$ is a basic sequence in $(\Lambda_1(\alpha), ||\cdot||_k)^{\wedge}$. Since this holds $\forall k$, $(y_n)$ is a basic sequence in $\Lambda_1(\alpha)$. ∎

(4.3) Corollary

If $\alpha_n = A^n$, $A > 1$ then $\Lambda_\infty(\alpha)$ is isomorphic to a subspace of $\Lambda_1(\alpha)$.

<u>Proof</u>

It is easy to verify the conditions of (4.2) with $b^k_n = e^{k\alpha_n}$, $M = 1$ and $A = A$. ∎

(4.4) Proposition

A finite type power series space $\Lambda_1(\beta)$ is isomorphic to a subspace of $\Lambda_1((2^n))$ iff (10) holds with $\alpha_n = 2^n$.

Proof

One way is clear so we suppose that $\Lambda_1(\beta)$ is isomorphic to a subspace of $\Lambda_1(\alpha)$, $\alpha_n = 2^n$. From I (6.2.2), (6.3.2) it follows that $\sup_n \frac{\alpha_n}{\beta_n} < \infty$, so $\Lambda_1(\beta)$ is not changed if we adjust $\beta \ni \alpha_n \leq \beta_n$ and $\log_2\beta_n$ is an integer. Hence it follows that we have strictly increasing sequences of indices $(r_n)$, $(\mu_n) \ni r_0 = 0$ and

$$\beta_m = 2^{\mu_n} = \alpha_{\mu_n} \text{ if } r_{n-1} < m \leq r_n, \quad n = 1,2,\ldots$$

and it follows that $\alpha_{r_n} \leq \beta_{r_n} = \alpha_{\mu_n}$ so $r_n \leq \mu_n \ \forall n$.

Let $y_n = \sum_i t_i^n e_i \in \Lambda_1(\alpha)$ be the image of $e_n$ in $\Lambda_1(\beta)$ as a subspace of $\Lambda_1(\alpha)$. Hence we can set

$$a_n^k = e^{-\frac{1}{\gamma_k}\alpha_n}, \quad b_n^k = |t_{q_n}^n{}_k| \, |a_{q_n}^k{}_k|, \quad c_n^k = e^{-\frac{1}{\rho_k}\beta_n} \quad \forall \, n,k$$

where $(\gamma_k)$, $(\rho_k)$ are increasing unbounded sequences chosen $\ni$

$$c_n^k \leq b_n^k \leq c_n^{k+1} \quad \forall k \text{ and } n \text{ sufficiently large.}$$

Let $q$, $j$, $k$, $\ell$ be indices, $r_{n-1} < m \leq r_n$. Then for $n$ sufficiently large we have,

$$\frac{|t_q^m|\,|a_q^j|}{|t_{q_m^k}^m|\,|a_{q_m^k}^k|} = \frac{|t_q^m|\,|a_q^\ell|}{|t_{q_m^k}^m|\,|a_{q_m^k}^k|}\,\frac{a_q^j}{a_q^\ell} \le \frac{b_m^\ell}{b_m^k}\,\frac{a_q^j}{a_q^\ell} \le \frac{c_m^{\ell+1}}{c_m^k}\,\frac{a_q^j}{a_q^\ell} =$$

$$e^{(\frac{1}{\rho_k} - \frac{1}{\rho_{\ell+1}})\alpha_{\mu_n} + (\frac{1}{\gamma_\ell} - \frac{1}{\gamma_j})\alpha_g}$$

Now let $\nu = \nu_{kj}$ be a positive integer sufficiently large $\ni$

$$|\frac{1}{\rho_k} - \frac{1}{\rho_{j+2}}| - (\frac{1}{\gamma_j} - \frac{1}{\gamma_{j+1}})2^\nu < -\frac{1}{2}$$

and

$$|\frac{1}{\gamma_{k-2}} - \frac{1}{\gamma_j}|2^{-\nu} < \frac{1}{2}(\frac{1}{\rho_{k-1}} - \frac{1}{\rho_k}) .$$

Then if $\mu_n + \nu < q$ we apply the above inequality with $\ell = j + 1$ so

$$\frac{|t_q^m|\,|a_q^j|}{|t_{q_m^k}^m|\,|a_{q_m^k}^k|} \le e^{|\frac{1}{\rho_k} - \frac{1}{\rho_{j+2}}|2^{\mu_n} - (\frac{1}{\gamma_j} - \frac{1}{\gamma_{j+1}})2^{\mu_n+\nu}} < e^{-\frac{1}{2}2^{\mu_n}} = e^{-\frac{1}{2}\alpha_{\mu_n}}$$

and if $q + \nu < \mu_n$ we apply it with $\ell = k - 2$ to obtain,

$$\frac{|t_q^m|\,|a_q^j|}{|t_{q_m^k}^m|\,|a_{q_m^k}^k|} \le e^{-(\frac{1}{\rho_{k-1}} - \frac{1}{\rho_k})2^{\mu_n} + |\frac{1}{\gamma_{k-2}} - \frac{1}{\gamma_j}|2^{\mu_n-\nu}} < e^{-\frac{1}{2}(\frac{1}{\rho_{k-1}} - \frac{1}{\rho_k})\alpha_{\mu_n}} ,$$

for n sufficiently large. Hence we have $\forall k,j \ \exists \ n_{kj}$ and $\delta_k > 0 \ni$

$$\frac{|t_q^m|a_q^j}{|t_{q_m^k}^m|a_{q_m^k}^k} < e^{-\delta_k \alpha_{\mu_n}} \text{ for } r_{n-1} < m \leq r_n,\ |\mu_n - q| > \nu_{kj}$$

$$\text{and } n \geq n_{kj} \quad .$$

Since $(y_n)$ is a basic sequence in $\Lambda_1(\alpha)$ this space has a fundamental sequence of norms $(|\cdot|_k) \ni (y_n)$ is a basic sequence in each $(\Lambda_1(\alpha), |\cdot|_k)^\wedge$ and we have indices k,j and $C > 0 \ni$ if $(||\cdot||_k)$ is the sequence of sup norms in $\Lambda_1(\alpha)$,

$$\frac{1}{C}||x||_3 \leq |x|_k \leq C||x||_j, \quad x \in \Lambda_1(\alpha).$$

Let $\nu = \nu_{3j}$, $\delta = \delta_3$ and define $(z_m)$ in $\Lambda_1(\alpha)$ by

$$z_m = \sum_{i=\mu_n-\nu}^{\mu_n+\nu} t_i^m e_i \text{ for } r_{n-1} < m \leq r_n$$

so by the above inequality,

$$\sum_m \frac{|y_m - z_m|_k}{|y_m|_k} \leq C^2 \sum \frac{||y_m - z_m||_j}{||y_m||_3} = C^2 \sum_n \sum_{m=r_{n-1}+1}^{r_n} \frac{\sup_{|\mu_n - q| > \nu} |t_q^m|a_q^j}{|t_{q_m^3}^m|a_{q_m^3}^3}$$

$$\leq BC^2 \sum_n (r_n - r_{n-1}) e^{-\delta\alpha_{\mu_n}} \leq BC^2 \sum_n \mu_n e^{-\delta 2^{\mu_n}} < \infty$$

where B comes from the fact that the estimate only holds for n sufficiently large and the series converges because $\mu_n < \mu_{n+1}$ so $n \leq \mu_n$.

Hence, as in (3.4) $\exists\ n_0 \ni (z_m)_{m>r_{n_0}}$ is a basic sequence.

In particular, it is linearly independent. Since $\alpha$ is weakly stable it suffices to establish (10) with $\beta$ replaced by $(\beta_m)_{m>r_{n_0}}$. To simplify notation we will assume that $n_0 = 0$ so $(z_m)$ is linearly independent.

If $0 \leq j < n$ then $\{z_m: r_j < m \leq r_n\}$ is a set of $r_n - r_j$ vectors contained in a $\mu_n - \mu_{j+1} + 2\nu + 1$ dimensional subspace of $\Lambda_1(\alpha)$ so

$$r_n - r_j \leq \mu_n - \mu_{j+1} + 2\nu + 1 \quad . \tag{12}$$

We will establish an injection of $\beta$ into $\alpha$ in which, for each n, the numbers $\beta_{r_{n-1}+1},\dots,\beta_{r_n}$ correspond to numbers in the set $\alpha_{\mu_n},\alpha_{\mu_n+1},\dots,\alpha_{\mu_n+2\nu}$. Since these $\beta_m$ are all equal to $\alpha_{\mu_n}$ and $\alpha$ is weakly stable, this will satisfy (10). Our only difficulty is that the sets $\{\alpha_m: \mu_n \leq m \leq \mu_n + 2\nu\}$ might overlap as n increases.

For n = 1, by (12) we can correspond $\beta_1,\dots,\beta_{r_1}$ to $\alpha_{\mu_1},\dots,\alpha_{\mu_1+r_1-1}$. Suppose that we have made the correspondence for $1,\dots,n-1$ in such a way that $\beta_m$ corresponds to $\alpha_i$ with i as small as possible consistent with the other requirements. It suffices to show that $\alpha_{\mu_n},\dots,\alpha_{\mu_n+2\nu}$ contains at least $r_n - r_{n-1}$ terms to which no $\beta_m$ has corresponded.

Let $\ell$ be the largest index $\leq$ n-1 so that in corresponding $\beta_{r_{\ell-1}+1},\dots,\beta_{r_\ell}$ the entire interval $\alpha_{\mu_\ell},\dots,\alpha_{\mu_\ell+2\nu}$ was available.

Such an index exists. For example, $\ell=1$. We then compute which $\alpha_i$ have been used up in making the correspondence $\beta_{r_{\ell-1}+1},\ldots,\beta_{r_{n-1}}$. By assumption, there are no gaps and we begin with $\alpha_{\mu_\ell}$ so that by (12) the number remaining is

$$\mu_n + 2\nu - \mu_\ell + 1 - (r_{n-1}-r_{\ell-1}) \geq r_n - r_{\ell-1} - (r_{n-1}-r_{\ell-1}) = r_n - r_{n-1},$$

and by definition of $\ell$ these are all in the set $\alpha_{\mu_n},\ldots,\alpha_{\mu_n+2\nu}$ so the induction is complete and the bijection is made. ∎

(4.5) It is interesting to compare (4.4) with (2.4.2), (3.6), (3.7) In all cases the condition is the same. It appears that (8) in (2.4.2) is weaker than (10) which is used in the others but it is easy to show that when $\alpha$ is stable, (8) implies (10) so they are equivalent.

(4.6) Proposition

Let $\alpha_n = 2^n$. Then no block basic sequence with respect to any basis in $\Lambda_1(\alpha)$ generates a subspace isomorphic to $\Lambda_\infty(\alpha)$.

<u>Proof</u>

Suppose we have such a block basic sequence which, by the quasi-equivalence property for $\Lambda_1(\alpha)$ we may assume to be with respect to a permutation of the coordinate basis in $\Lambda_1(\alpha)$. By the quasi-equivalence property for $\Lambda_\infty(\alpha)$ we may assume that the block basic sequence $(y_n)$ is in fact the image of the coordinate basis in $\Lambda_\infty(\alpha)$ as a subspace of $\Lambda_1(\alpha)$.

Thus we have a decomposition $\mathbb{N} = \bigcup_n \mathbb{N}_n$ of $\mathbb{N}$ into pairwise disjoint finite sets $\mathbb{N}_n$ each of which we assume to be arranged in increasing order and

$$y_n = \sum_{i \in \mathbb{N}_n} t_i e_i, \quad n = 1,2,\dots .$$

We can set

$$a_n^k = e^{-\frac{1}{\gamma_k}\alpha_n}, \quad b_n^k = |t_{q_n^k}| a_{q_n^k}^k, \quad c_n^k = e^{\rho_k \alpha_n}$$

where $(\gamma_k)$, $(\rho_k)$ are increasing unbounded sequences chosen ∋

$$c_n^k \le b_n^k \le c_n^{k+1} \text{ for n sufficiently large.}$$

Applying (1.3),

$$e^{(\frac{1}{\gamma_k} - \frac{1}{\gamma_{k+1}})\alpha_{q_n^k}} = \frac{a_{q_n^k}^{k+1}}{a_{q_n^k}^k} \le \frac{b_n^{k+1}}{b_n^k} \le \frac{c_n^{k+2}}{c_n^k} = e^{(\rho_{k+2}-\rho_k)\alpha_n}$$

n sufficiently large

and using the fact that $\alpha_n = 2^n$, we have integers $\nu_k > 0$ ∋ $\nu_k < \nu_{k+1}$ and

$$q_n^k \le n + \nu_k \quad \forall\, n, k .$$

Next we will show that ∃ $n_0$ ∋ $\sup_{n \ge n_0} \max\{k : q_n^k = q_n^1\} = N < \infty$.

If not, then ∃ indices $(n_j)$, $n_j < n_{j+1}$ ∋ $\max\{k : q_{n_j}^k = q_{n_j}^1\} \ge j$ so

$$b_{n_j}^k = |t_{q_{n_j}^k}| a_{q_{n_j}^k}^k = |t_{q_{n_j}^1}| e^{-\frac{1}{\gamma_k}\alpha_{q_{n_j}^1}} \qquad k \le j$$

which implies that $(y_n)$ has a subsequence which generates a subspace isomorphic to a finite type power series space which is impossible (say by II (1.1) and (1.2)) which proves the claim.

Let $\mu_1 = n_0 + \nu_{N+1} - 1$, $\mu_0 = \mu_1 - (\nu_{N+1}-\nu_1)$.

Now $1 \leq q_n^1 \leq n + \nu_1$ and $q_n^{N+1} \neq q_n^1$ for $n \geq n_0$ whereas, if $n \leq n_0 + \mu_0$,

$$1 \leq q_n^{N+1} \leq n + \nu_{N+1} \leq n_0 + \mu_0 + \nu_{n+1} = n_0 + \mu_1 + \nu_1$$

Since $q_n^k \in N_n$ $\forall n,k$ it follows that each of the sets $\mathbb{N}_n, n_0 \leq n \leq n_0 + \mu_0$ meets the set $\{1,2,\ldots,n_0+\mu_1+\nu_1\}$ in at least two distinct points $q_n^1$, $q_n^{N+1}$. Also, each set $N_n$, $n_0 + \mu_0 < n \leq n_0 + \mu_1$ meets this set in at least one point, $q_n^1$. Hence since $\{\mathbb{N}_n : n_0 \leq n \leq n_0 + \mu_1\}$ are pairwise disjoint, then

$$2(\mu_0+1) + \mu_1 - \mu_0 \leq \mu_1 + \nu_1$$

which is false so we have our contradiction.

■

## 5. Mixed Power Series Spaces

(5.1) It ought to be possible to say something about the subspaces of $\Lambda_1(\alpha)\times\Lambda_\infty(\beta)$ in terms of the subspaces of $\Lambda_1(\alpha)$, $\Lambda_\infty(\beta)$. Also, one could try to do the same thing for the tensor product. Actually there are no results in this direction and more study will be required.

In this section we set down, briefly, some preliminary observations in the hope of suggesting lines of investigation.

(5.2) If $\alpha$ is stable and (8) holds then the subspaces of $\Lambda_1(\alpha)\times\Lambda_\infty(\beta)$ are the same as the subspaces of $\Lambda_1(\alpha)$. Indeed one way is clear and for the other we conclude from (2.4.4) and the stability that $\Lambda_1(\alpha)\times\Lambda_\infty(\beta)$ is isomorphic to a subspace of $\Lambda_1(\alpha)\times\Lambda_1(\alpha) = \Lambda_1(\alpha)$.

(5.3) If $(z_n)$ is a basic sequence in $\Lambda_1(\alpha)\times\Lambda_\infty(\beta)$ which is a block basic sequence with respect to some permutation of the standard basis $e_n^1 + e_n^\infty$ where $(e_n^1)$, $(e_n^\infty)$ are the coordinate bases in $\Lambda_1(\alpha)$, $\Lambda_\infty(\beta)$ respectively, then the subspace generated by $(z_n)$ is isomorphic to some space $K(a+b)$, $a + b = (a_n^k + b_n^k)$ where $K(a)$, $K(b)$ are isomorphic to subspaces of $\Lambda_1(\alpha)$, $\Lambda_\infty(\beta)$ respectively.

To see this we note that there are bijections $\sigma,\tau$: $\mathbb{N}\times\mathbb{N}\to\mathbb{N}$ so that we can write

$$z_n = \sum_i (s_{\sigma(i,n)} e^1_{\sigma(i,n)} + t_{\tau(i,n)} e^\infty_{\tau(i,n)}) \quad .$$

We then take

$$a_n^k = ||\sum_i s_{\sigma(i,n)} e^1_{\sigma(i,n)}||_k, \quad b_n^k = ||\sum_i t_{\tau(i,n)} e^\infty_{\tau(i,n)}||_k$$

where $||\cdot||_k$ is the "sup norm" in $\Lambda_1(\alpha)$, $\Lambda_\infty(\beta)$.

## 6. $L_f(\alpha,r)$ spaces with $r = 0,\infty$

(6.1) In this section we assume that f is a Dragilev function and that $\alpha$ satisfies the stability condition (2.1). Again this is equivalent by the quasi-equivalence property to the fact that $L_f(\alpha,r) \cong L_f(\alpha,r)\times L_f(\alpha,r)$.

We will not discuss the cases $r < 0$, $0 < r < \infty$ in these notes because there are some difficulties in obtaining what seems to be the natural results. One obvious difference is that it is no longer the case that the stability condition is equivalent to the space being isomorphic to its cartesian square. There are other difficulties.

(6.2) Characterizations

(6.2.1) Theorem

Let F be a nuclear Fréchet space with basis $(y_n)$. Then F is isomorphic to a subspace of $L_f(\alpha,\infty)$ iff

i) The basis $(y_n)$ has a representation b $\ni \exists M > 1 \ni$

$$f^{-1}\left(\log\frac{b_n^{k+1}}{b_n^k}\right) \leq \frac{1}{M} f^{-1}\left(\log\frac{b_n^{k+2}}{b_n^{k+1}}\right) \quad \forall\ n,k$$

ii) $\forall$ nbd of 0 U in F and $\rho > 0$ $\exists$ nbd of 0 V in F $\ni$

$$d_n(V,U) \leq e^{-f(\rho\alpha_n)} \quad n \text{ sufficiently large .}$$

<u>Proof</u>

First we show that the conditions are necessary. It is easy to check that ii) holds for $L_f(\alpha,\infty)$ so by I (6.2.4) it holds for F.

For i) we apply (1.3) with $a_n^k = e^{f(\rho_k\alpha_n)}$ where $(\rho_k)$ is an increasing unbounded sequence chosen $\ni$ $f(2x) \geq 2f(x)$, $x \geq \rho_1\alpha_1$ and $\rho_{k+1} \geq 4\rho_k \geq 1$. Then

$$\frac{b_n^{k+1}}{b_n^k} \leq \frac{a_{q_n^{k+1}}^{k+1}}{a_{q_n^{k+1}}^k}, \quad \frac{a_{q_n^{k+1}}^{k+2}}{a_{q_n^{k+1}}^{k+1}} \leq \frac{b_n^{k+2}}{b_n^{k+1}} \qquad n,k$$

and so, since $\alpha_{q_n^{k+1}} \geq \alpha_1$,

$$f^{-1}\left(\log\frac{b_n^{k+1}}{b_n^k}\right) \le f^{-1}\left(f(\rho_{k+1}\alpha_{q_n^{k+1}}) - f(\rho_k\alpha_{q_n^{k+1}})\right) \le \rho_{k+1}\alpha_{q_n^{k+1}} \le$$

$$\le \frac{1}{4}\rho_{k+2}\alpha_{q_n^{k+1}} = \frac{1}{2}f^{-1}\circ f\left(\frac{1}{2}\rho_{k+2}\alpha_{q_n^{k+1}}\right) \le \frac{1}{2}f^{-1}\left(\frac{1}{2}f(\rho_{k+2}\alpha_{q_n^{k+1}})\right)$$

$$\le \frac{1}{2}f^{-1}\left(f(\rho_{k+2}\alpha_{q_n^{k+1}}) - f(\rho_{k+1}\alpha_{q_n^{k+1}})\right) \le \frac{1}{2}f^{-1}\left(\log\frac{b_n^{k+2}}{b_n^{k+1}}\right)$$

which is condition i) with $M = 2$.

Now we assume that i) and ii) hold and we will construct a basic sequence $(z_m)$ in $L_f(\alpha,\infty)$ and a matrix b which is a representation of $(z_m)$ and of a permutation of $(y_n)$. This will complete the proof.

Let c be a representation of $(y_n)$ satisfying i). Then by ii) and the basic properties of diameters we have a permutation $\pi$ of $\mathbb{N}$ and $j \in \mathbb{N} \ni$

$$\frac{c^1_{\pi(n)}}{c^j_{\pi(n)}} \le e^{-f(A\alpha_n)} \quad \text{for n sufficiently large .}$$

(Here A is given by (2.1)). We can easily adjust c so that this relation holds $\forall$ n and i) is still true. Now let $(j_k)$ be any subsequence of indices. Then if R is given by I (5.4), it follows from i) (used first with M replaced by 1 and then with M) that

$$f^{-1}\left(\log\frac{c_n^{j_{k+1}}}{c_n^{j_k}}\right) = f^{-1}\left(\log\frac{c_n^{j_{k+1}}}{c_n^{j_{k+1}-1}} + \ldots + \log\frac{c_n^{j_{k+1}}}{c_n^{j_k}}\right) \le f^{-1}\left((j_{k+1}-j_k)\log\frac{c_n^{j_{k+1}}}{c_n^{j_{k+1}-1}}\right)$$

$$\leq R(j_{k+1}-j_k) f^{-1}(\log\frac{c_n^{j_{k+1}}}{c_n^{j_{k+1}-1}}) \leq \frac{R(j_{k+1}-j_k)}{M^{j_{k+2}-j_{k+1}}} f^{-1}(\log\frac{c_n^{j_{k+2}}}{c_n^{j_{k+1}}}) .$$

We then choose $(j_k)$ so that if $\rho_1 = 1$, $\rho_2 \geq j$ and

$$\rho_{k+1} \leq \frac{M^{j_{k+2}-j_{k+1}}}{2AR^2(j_{k+1}-j_k)} \rho_k \quad \forall k$$

then $(\rho_k)$ is an increasing unbounded sequence and if we set

$b_n^k = c_{\pi(n)}^{j_k}$ then b is a representation of a permutation of $(y_n) \ni$

$$\frac{b_n^1}{b_n^2} \leq e^{-f(A\alpha_n)} \leq e^{-f(\alpha_{2n-1})} \quad \forall n \tag{13}$$

and

$$f^{-1}(\log\frac{b_n^{k+1}}{b_n^k}) \leq \frac{\rho_k}{2AR\rho_{k+1}} f^{-1}(\log\frac{b_n^{k+2}}{b_n^{k+1}}) \quad \forall n,k . \tag{14}$$

(Compare this with II (1.5.1)).

Now we fix m and generally suppress it in our notation. Let $\rho_0 = 0$, $x_\nu = e_{2^{\nu-1}(2m-1)}$, $\beta_\nu = \alpha_{2^{\nu-1}(2m-1)}$, $a_\nu^k = e^{f(\rho_{k-1}\beta_\nu)}$,

$k,\nu \in \mathbb{N}$ so that $(x_\nu)$ is a subsequence of $(e_n)$ (with no repetitions as m varies) and $(a_\nu^k)$ is a strict type $(d_0)$ representation of $(x_\nu)$. We will select $q^k$, $t_{q^k}$ to define

$$z_m = \sum_{k=1}^{m} t_{q^k} x_{q^k} \quad .$$

First we show that we can select $q^1,\ldots,q^k$ ∋ for $k = 1,\ldots,$ $m-1$,

$$f(\rho_k\beta_{q^k})-f(\rho_{k-1}\beta_{q^k}) = \log\frac{a^{k+1}_{q^k}}{a^k_{q^k}} \le \log\frac{b^{k+1}_m}{b^k_m} < \log\frac{a^{k+1}_{q^{k+1}}}{a^k_{q^{k+1}}}$$

$$= f(\rho_k\ \beta_{q^{k+1}}) - f(\rho_{k-1}\beta_{q^{k+1}}) \ .$$

By (13) we can choose $q^1 = 1$ and the left inequality holds. Suppose that we have chosen $q^1,\ldots,q^k$ ∋ the left inequality and also the right inequality holds. Since $\lim_\nu \beta_\nu = \infty$ we can let $q^{k+1}$ be the smallest index ∋ the right inequality holds. Then by I (5.3) we have, for m sufficiently large,

$$\frac{1}{2}f(\rho_k\beta_{q^{k+1}-1}) \le f(\rho_k\beta_{q^{k+1}-1}) - f(\rho_{k-1}\beta_{q^{k+1}-1}) \le \log\frac{b^{k+1}_m}{b^k_m}$$

so

$$\beta_{q^{k+1}-1} \le \frac{1}{\rho_k}f^{-1}(2\log\frac{b^{k+1}_m}{b^k_m}) \le \frac{2R}{\rho_k}f^{-1}(\log\frac{b^{k+1}_m}{b^k_m})$$

so by (14) and the fact that f is increasing,

$$f(\rho_{k+1}\beta_{q^{k+1}}) - f(\rho_k\beta_{q^{k+1}}) \le f(\rho_{k+1}\beta_{q^{k+1}}) \le f(A\rho_{k+1}\beta_{q^{k+1}-1})$$

$$\le f(\frac{2AR\rho_{k+1}}{\rho_k}f^{-1}(\log\frac{b^{k+1}_m}{b^k_m})) \le \log\frac{b^{k+2}_m}{b^{k+1}_m} \text{ for m sufficiently large ,}$$

so the construction of $q^1,\ldots,q^m$ is complete.

Now set

$$t_{q^k} = \frac{b_m^k}{a_{q^k}^k} \qquad k = 1,\ldots,m$$

and it follows from our construction that the condition of II (2.2.3) is satisfied so it follows that we have $m_0 \ni (z_m)_{m\geq m_0}$ is a basic sequence in $L_f(\alpha,\infty)$ for which $(b_m^k)_{k,m\geq m_0}$ is a representation. But $(z_m)$ is also a basic sequence in $L_f(\alpha,\infty)$ and b is a representation for it.

∎

(6.2.2) Theorem

Let F be a nuclear Fréchet space with basis $(y_n)$. Then F is isomorphic to a subspace of $L_f(\alpha,0)$ iff

i) The basis $(y_n)$ has a representation $b \ni \exists\, M > 1 \ni$

$$f^{-1}\left(\log\frac{b_n^{k+1}}{b_n^k}\right) \leq M f^{-1}\left(\log\frac{b_n^{k+2}}{b_n^{k+1}}\right) \quad \forall\, n,k$$

ii) $\forall$ nbd of 0 U in F $\exists$ nbd of 0 V in F and $\rho > 0 \ni$

$$d_n(V,U) \leq e^{-f(\rho\alpha_n)} \quad n \text{ sufficiently large}.$$

<u>Proof</u>

First we show that the conditions are necessary. It is easy to check that ii) holds for $L_f(\alpha,0)$ so by I (6.2.4) it holds for F.

For i) we apply (1.3) with $a_n^k = e^{-f(\frac{1}{\rho_k}\alpha_n)}$ where $\rho_{k+1} = 2^k$.

Then by the nuclearity of F it follows from (1.3) that

$\exists\, j_0 \ni \lim_n q_n^k = \infty \quad \forall k \geq j_0$. Also we have,

$$\frac{b_n^{k+1}}{b_n^k} \leq \frac{a_{q_n^{k+1}}^{k+1}}{a_{q_n^{k+1}}^{k}}, \qquad \frac{a_{q_n^{k+1}}^{k+2}}{a_{q_n^{k+1}}^{k+1}} \leq \frac{b_n^{k+2}}{b_n^{k+1}} \quad \forall n,\ k \ .$$

Hence, for $k \geq k_0$ and n sufficiently large we have by I (5.3),

$$f^{-1}(\log\frac{b_n^{k+1}}{b_n^k}) \leq f^{-1}(f(\frac{1}{\rho_k}\alpha_{q_n^{k+1}})-f(\frac{1}{\rho_{k+1}}\alpha_{q_n^{k+1}})) \leq \frac{1}{\rho_k}\alpha_{q_n^{k+1}} \leq \frac{2}{\rho_{k+1}}\alpha_{q_n^{k+1}}$$

$$= 4f^{-1}\circ f(\frac{1}{2\rho_{k+1}}\alpha_{q_n^{k+1}}) \leq 4f^{-1}(f(\frac{1}{\rho_{k+1}}\alpha_{q_n^{k+1}})-f(\frac{1}{\rho_{k+2}}\alpha_{q_n^{k+1}}))$$

$$\leq 4f^{-1}(\log\frac{b_n^{k+2}}{b_n^{k+1}}) \ .$$

This gives i) for $k \geq k_0$ and n sufficiently large and we can clearly adjust b so as to obtain i) for all n,k.

Now we assume that i) and ii) hold and we will construct a basic sequence $(z_m)$ in $L_f(\alpha,0)$ and a matrix b which is a representation of $(z_m)$ and of a permutation of $(y_n)$. This will complete the proof.

Let c be a representation of $(y_n)$ satisfying i). Then by ii) and the basic properties of diameters we have a permutation $\pi$ of $\mathbb{N}$, and index j and $\rho > 0 \ni$

$$\frac{c^1_{\pi(n)}}{c^{j+1}_{\pi(n)}} \le e^{-f(\rho\alpha_n)} \qquad \text{for } n \text{ sufficiently large}$$

By decreasing $\rho$ if necessary we can be sure that this holds $\forall n$.

Now let $b = (b^k_n)$ be given by $b^k_n = (c^{(k-1)j+1}_{\pi(n)})$. Then for some $\ell \ni (k-1)j < \ell \le kj$, writing $m = \pi(n)$,

$$f^{-1}(\log\frac{b^{k+1}_n}{b^k_n}) = f^{-1}(\log\frac{c^{kj+1}_m}{c^{(k-1)j+1}_m}) = f^{-1}(\log\frac{c^{kj+1}_m}{c^{kj}_m}+\ldots+\log\frac{c^{(k-1)j+2}_m}{c^{(k-1)j+1}_m})$$

$$\le f^{-1}(j\log\frac{c^{\ell+1}_m}{c^{\ell}_m}) \le Rjf^{-1}(\log\frac{c^{\ell+1}_m}{c^{\ell}_m}) \le M_k f^{-1}(\log\frac{c^{kj+2}_m}{c^{kj+1}_m})$$

$$\le M_k f^{-1}(\log\frac{c^{(k+1)j+1}_m}{c^{kj+1}_m}) = M_k f^{-1}(\log\frac{b^{k+2}_n}{b^{k+1}_n}) \quad ,$$

where $M_k > 0$ is appropriately chosen, and we have applied I (5.4).

Thus we have a representation b of a permutation of $(y_n)$ $\rho > 0$ and a sequence $(M_k)$ of positive numbers $\ni$

$$\frac{b^1_n}{b^2_n} \le e^{-f(\rho\alpha_n)} \le e^{-f(\frac{\rho}{A}\alpha_{2n-1})} \qquad \forall n \tag{15}$$

and

$$f^{-1}(\log\frac{b^{k+1}_n}{b^k_n}) \le M_k f^{-1}(\log\frac{b^{k+2}_n}{b^{k+1}_n}) \qquad \forall n,k \tag{16}$$

Now we fix m and generally suppress it in our notation. Let $(\rho_k)$ be a strictly increasing sequence $\ni$ $\rho_1 \ge \frac{A}{\rho}$ and $\rho_{k+1} \ge$

$2ARM_k\rho_k \; \forall k$. Let $x_\nu = e_{2^{\nu-1}(2m-1)}$, $\beta_\nu = \alpha_{2^{\nu-1}(2m-1)}$, $a_\nu^k = e^{-f(\frac{1}{\rho_k}\beta_\nu)}$, $k, \nu \in \mathbb{N}$ so that $(x_\nu)$ is a subsequence of $(e_n)$ (with no repetitions as m varies) and $(a_\nu^k)$ is a strict type $(d_0)$ representation of $(x_\nu)$. We will select $q^k$, $t_{q^k}$ to define

$$z_m = \sum_{k=1}^{m} t_{q^k} x_{q^k} \quad .$$

First we show that we can select $q^1,\ldots,q^k \ni$ for $k = 1,\ldots,$ $m-1$,

$$f(\frac{1}{\rho_k}\beta_{q^k}) - f(\frac{1}{\rho_{k+1}}\beta_{q^k}) = \log\frac{a_{q^k}^{k+1}}{a_{q^k}^{k}} \le \log\frac{b_m^{k+1}}{b_m^k}$$

$$< \log\frac{a_{q^{k+1}}^{k+1}}{a_{q^{k+1}}^{k}} = f(\frac{1}{\rho_k}\beta_{q^{k+1}}) - f(\frac{1}{\rho_{k+1}}\beta_{q^{k+1}}) \quad .$$

By (15) we can choose $q^1 = 1$ and the left inequality holds. Suppose that we have chosen $q^1,\ldots,q^k \ni$ the left inequality and also the right inequality holds. Since $\lim_\nu \beta_\nu = \infty$ we can let $q^{k+1}$ be the smallest index $\ni$ the right inequality holds. Then by I (5.3) we have, for m sufficiently large,

$$\tfrac{1}{2}f(\frac{1}{\rho_k}\beta_{q^{k+1}-1}) \le f(\frac{1}{\rho_k}\beta_{q^{k+1}-1}) - f(\frac{1}{\rho_{k+1}}\beta_{q^{k+1}-1}) \le \log\frac{b_m^{k+1}}{b_m^k}$$

so by (16) and I (5.4) we have for m sufficiently large,

$$f(\frac{1}{\rho_{k+1}}\beta_{q^{k+1}}) - f(\frac{1}{\rho_{k+2}}\beta_{q^{k+2}}) \leq f(\frac{1}{\rho_{k+1}}\beta_{q^{k+1}}) \leq f(\frac{A}{\rho_{k+1}}\beta_{q^{k+1}-1})$$

$$= f(\frac{A\rho_k}{\rho_{k+1}}\frac{1}{\rho_k}\beta_{q^{k+1}-1}) \leq f(\frac{A\rho_k}{\rho_{k+1}} f^{-1}(2\log\frac{b_m^{k+1}}{b_m^k})) \leq f(\frac{2AR\rho_k}{\rho_{k+1}} f^{-1}(\log\frac{b_m^{k+1}}{b_m^k}))$$

$$\leq f(\frac{2ARM_k\rho_k}{\rho_{k+1}} f^{-1}(\log\frac{b_m^{k+2}}{b_m^{k+1}})) \leq \log\frac{b_m^{k+2}}{b_m^{k+1}} ,$$

so the construction of $q^1,\ldots,q^m$ is complete.

Now set

$$t_{q^k} = \frac{b_m^k}{a_{q^k}^k}, \quad k = 1,\ldots,m$$

and it follows from our construction that the condition of II (2.2.3) is satisfied so we have $m_0 \ni (z_m)_{m\geq m_0}$ is a basic sequence in $L_f(\alpha,0)$ for which $(b_m^k)_{k,m\geq m_0}$ is a representation. But $(z_m)$ is also a basic sequence in $L_f(\alpha,0)$ and b is a representation for it. ∎

(6.3) Power series subspaces of $L_f(\alpha,r)$ spaces, $r = 0,\infty$

(6.3.1) Proposition

No space $\Lambda_\infty(\beta)$ is isomorphic to a subspace of $L_f(\alpha,\infty)$.

Proof

We will show that $\Lambda_\infty(\beta)$ does not satisfy condition i) of (6.2.1). If it did, then we would have a matrix b satisfying that relation and also subsequences of indices $(j_k)$ and $(\ell_k) \ni$

$$b_n^{j_k} \leq e^{\ell_k\beta_n} \leq b_n^{j_{k+1}} \quad \text{n sufficiently large}$$

and so, applying ii) (twice) and I (5.4) we have $M_k > 1$ for k sufficiently large ∋

$$f^{-1}((\ell_{k+1}-\ell_k)\beta_n) \le f^{-1}\left(\log\frac{b_n^{j_{k+2}}}{b_n^{j_k}}\right) = f^{-1}\left(\log\frac{b_n^{j_{k+2}}}{b_n^{j_{k+2}-1}} + \ldots + \log\frac{b_n^{j_k+1}}{b_n^{j_k}}\right)$$

$$\le R(j_{k+2}-j_k)f^{-1}\left(\log\frac{b_n^{j_{k+2}}}{b_n^{j_{k+2}-1}}\right) \le \frac{1}{M_k}f^{-1}\left(\log\frac{b_n^{j_{k+4}}}{b_n^{j_{k+2}}}\right) \le \frac{1}{M_k}f^{-1}((\ell_{k+4}-\ell_{k+1})\beta_n)$$

for n sufficiently large. But, since $\lim_n \beta_n = \infty$, this violates I (5.3) iii).

∎

(6.3.2) Proposition

A space $\Lambda_1(\beta)$ is isomorphic to a subspace of $L_f(\alpha,0)$ iff

$$\inf_{\rho>0} \overline{\lim_n} \frac{f(\rho\alpha_n)}{\beta_n} = 0$$

Proof

If we take $b_n^k = e^{-\frac{1}{2^k}\beta_n}$ then condition i) of (6.2.2) follows from I (5.4) with M = 2R. It is clear that condition ii) of (6.2.2) is equivalent to the given condition.

∎

(6.3.3) Proposition

A space $\Lambda_\infty(\beta)$ is isomorphic to a subspace of $L_f(\alpha,0)$ iff

$$\inf_{\rho>0} \overline{\lim_n} \frac{f(\rho\alpha_n)}{\beta_n} = 0$$

Proof

Any space with a basis of type $(d_3)$ (such as $\Lambda_\infty(\beta)$) clearly satisfies condition i) of (6.2.2) with $M = 1$. It is clear that condition ii) of (6.2.2) is equivalent to the given condition.

∎

(6.3.4) Proposition

No space $\Lambda_1(\beta)$ is isomorphic to a subspace of $L_f(\alpha,\infty)$.

Proof

The space $\Lambda_1(\beta)$ has a basis of type $(d_2)$ whereas it is easy to see that the coordinate basis in $L_f(\alpha,\infty)$ is type $(d_3)$ so the result follows from (2.3.3)

∎

(6.4) $L_g(\beta,s)$ subspaces of $L_f(\alpha,r)$, $r = 0,\infty$, $s = \pm 1, 0,\infty$.

(6.4.1) We have a complete characterization arising from (6.2.1), (6.2.2) along with computations quite similar to the ones we have been making. First we prove two lemmas analogous to the results in II (1.3).

(6.4.2) Lemma

If F is a nuclear Fréchet space with a basis $(y_n)$ having a representation b satisfying (6.2.1) i) and c is any representation of $(y_n)$, then $\exists$ a subsequence of indices $(j_k)$ and $M_1 > 1 \ni$ $\forall k \ \exists \ n_k \ni$

$$f^{-1}\left(\log\frac{c_n^{j_{k+1}}}{c_n^{j_k}}\right) \le \frac{1}{M_1} f^{-1}\left(\log\frac{c_n^{j_{k+2}}}{c_n^{j_{k+1}}}\right), \quad n \ge n_k .$$

Proof

We have $(j_k)$ and also a subsequence $(\ell_n) \ni$ for n sufficiently large,

$$c_n^{j_k} \le b_n^{\ell_k} \le b_n^{\ell_k+1} \le c_n^{j_{k+1}} .$$

Let $1 < M_1 < M$ so by (6.2.1) i) and I (5.3) we have, for some $\ell \ni \ell_{k-1} \le \ell < \ell_{k+1}$

$$f^{-1}(\log\frac{c_n^{j_{k+1}}}{c_n^{j_k}}) \le f^{-1}(\log\frac{b_n^{\ell_{k+1}}}{b_n^{\ell_{k-1}}}) \le f^{-1}((\ell_{k+1}-\ell_{k-1})\log\frac{b_n^{\ell+1}}{b_n^{\ell}})$$

$$\le \frac{M}{M_1}f^{-1}(\log\frac{b_n^{\ell+1}}{b_n^{\ell}}) \le \frac{1}{M_1}f^{-1}(\log\frac{b_n^{\ell_{k+1}+1}}{b_n^{\ell_{k+1}}}) \le \frac{1}{M_1}f^{-1}(\log\frac{c_n^{j_{k+2}}}{c_n^{j_{k+1}}})$$

for n sufficiently large. ∎

(6.4.3) Lemma

If F is a nuclear Fréchet space with a basis $(y_n)$ having a representation b satisfying (6.2.2) i) and c is any representation of $(y_n)$, then ∃ a subsequence of indices $(j_k)$ and $M_1 > 1$ $\ni \forall k \,\exists n_k \ni$

$$f^{-1}(\log\frac{c_n^{j_{k+1}}}{c_n^{j_k}}) \le M_1 f^{-1}(\log\frac{c_n^{j_{k+2}}}{c_n^{j_{k+1}}}), \qquad n \ge n_k$$

Proof

We repeat the argument for (6.4.2) except that $M_1$ is chosen so that $M_1 > M$ and in the computation, $M$, $M_1$ are interchanged, ∎

(6.4.4) Proposition

$L_g(\beta,\infty)$ is isomorphic to a subspace of $L_f(\alpha,\infty)$ iff

i) $\exists M_1 > 1 \ni \forall \rho \geq 1 \exists \tau > \rho \ni \overline{\lim_n} \dfrac{f^{-1}g(\rho\beta_n)}{f^{-1}g(\tau\beta_n)} \leq \dfrac{1}{M_1}$

ii) $\forall \rho > 0 \exists \tau > 0 \ni \sup_n \dfrac{f(\rho\alpha_n)}{g(\tau\beta_n)} < \infty$ .

Proof

It is a straightforward computation to verify that ii) here is equivalent to (6.2.1) ii).

Suppose that i) holds. Then we can choose $(\rho_k)$ increasing to $\infty \ni$

$$\overline{\lim_n} \frac{f^{-1}g(\rho_k\beta_n)}{f^{-1}g(\rho_{k+1}\beta_n)} \leq \frac{1}{M_1} .$$

Then set $b_n^k = e^{g(\rho_k\beta_n)}$ and choose $M \ni 1 < M < M_1$. By I (5.3), (5.4) we have, for n sufficiently large,

$$f^{-1}(\log\frac{b_n^{k+1}}{b_n^k}) \leq f^{-1}g(\rho_{k+1}\beta_n) \leq \frac{1}{M_1}f^{-1}g(\rho_{k+2}\beta_n) \leq \frac{1}{M}f^{-1}(\tfrac{1}{2}g(\rho_{k+2}\beta_n))$$

$$\leq \frac{1}{M}f^{-1}(g(\rho_{k+2}\beta_n) - g(\rho_{k+1}\beta_n)) = \frac{1}{M}f^{-1}(\log\frac{b_n^{k+2}}{b_n^{k+1}})$$

which gives (6.2.1) i) for n sufficiently large so we can adjust finitely many values of n for each k to obtain a representation for which (6.2.1) ii) holds.

Finally, if (6.2.1) i) holds we set $c_n^k = e^{g(k\beta_n)}$ and invoke (6.4.2) to obtain,

$$\overline{\lim_n} \frac{f^{-1}(g(j_{k+1}\beta_n)-g(j_k\beta_n))}{f^{-1}(g(j_{k+2}\beta_n)-g(j_{k+1}\beta_n))} \le \frac{1}{M_1}$$

and from I (5.3) it follows that

$$\overline{\lim_n} \frac{f^{-1}(g(j_{k+1}\beta_n))}{f^{-1}(g(j_{k+2}\beta_n))} \le \frac{1}{M_1} \quad \forall k$$

from which i) follows. ∎

(6.4.5) Proposition

$L_g(\beta,1)$ is isomorphic to a subspace of $L_f(\alpha,\infty)$ iff

i) $\exists M_1 > 1 \ni \forall \rho \in (0,1) \ \exists \tau \in (0,1) \ni \overline{\lim_n} \frac{f^{-1}g(\rho\beta_n)}{f^{-1}g(\tau\beta_n)} \le \frac{1}{M_1}$

ii) $\forall \rho > 0 \ \exists \ \tau \in (0,1) \ni \sup_n \frac{f(\rho\alpha_n)}{g(\tau\beta_n)} < \infty$ .

Proof

We repeat the proof of (6.4.4) verbatim except that we choose $(\rho_k)$ increasing to 1; we set $c_n^k = e^{g(\frac{k}{k+1}\beta_n)}$ and the final computation is changed to

$$\overline{\lim_n} \frac{f^{-1}\left(g\left(\frac{j_{k+1}}{j_{k+1}+1}\beta_n\right) - g\left(\frac{j_k}{j_k+1}\beta_n\right)\right)}{f^{-1}\left(g\left(\frac{j_{k+2}}{j_{k+2}+1}\beta_n\right) - g\left(\frac{j_{k+1}}{j_{k+1}+1}\beta_n\right)\right)} \le \frac{1}{M_1}$$

and again from I (5.3) it follows that

$$\overline{\lim_n} \frac{f^{-1}(g(\frac{j_{k+1}}{j_{k+1}+1}\beta_n))}{f^{-1}(g(\frac{j_{k+2}}{j_{k+2}+1}\beta_n))} \leq \frac{1}{M_1}$$

from which i) follows. ∎

(6.4.6) Proposition

No space $L_g(\beta,0)$ is isomorphic to a subspace of $L_f(\alpha,\infty)$.

Proof

In this case, condition i) of (6.2.1) and (6.4.2) imply that for some $(j_k)$ and $M_1 > 1$ we have for n sufficiently large

$$f^{-1}(g(\frac{1}{j_k}\beta_n) - g(\frac{1}{j_{k+1}}\beta_n)) \leq \frac{1}{M_1}f^{-1}(g(\frac{1}{j_{k+1}}\beta_n) - g(\frac{1}{j_{k+2}}\beta_n))$$

and hence, by I (5.3),

$$1 \leq \overline{\lim_n} \frac{f^{-1}(g(\frac{1}{j_k}\beta_n))}{f^{-1}(g(\frac{1}{j_{k+1}}\beta_n))} < 1$$

which is a contradiction. ∎

(6.4.7) Proposition

No space $L_g(\beta,-1)$ is isomorphic to a subspace of $L_f(\alpha,\infty)$.

Proof

Again we apply (6.2.1) i) and (6.4.2) but this time with $c_n^k = e^{-f(\rho_k\beta_n)}$, where $(\rho_k)$ decreases to 1. We then have $(j_k)$

and $M_1 > 1 \ni$ for n sufficiently large,

$$f^{-1}(g(\rho_{j_k}\beta_n) - g(\rho_{j_{k+1}}\beta_n)) \leq \frac{1}{M_1} f^{-1}(g(\rho_{j_{k+1}}\beta_n) - g(\rho_{j_{k+2}}\beta_n))$$

and by I (5.3) we obtain

$$1 \leq \overline{\lim} \frac{f^{-1}(g(\rho_{j_k}\beta_n))}{f^{-1}(g(\rho_{j_{k+1}}\beta_n))} < 1$$

which is a contradiction. ∎

(6.4.8) Proposition

$L_g(\beta,\infty)$ is isomorphic to a subspace of $L_f(\alpha,0)$ iff

$$\exists\ \rho,\ \tau > 0 \ni \sup_n \frac{f(\rho\alpha_n)}{g(\tau\beta_n)} < \infty$$

Proof

It is straightforward to check that this condition is equivalent to (6.2.2) ii) and we now check that (6.2.2) i) is always satisfied. Set $b_n^k = e^{g(k\beta_n)}$ and since $f^{-1}g((k+1)\beta_n) \leq f^{-1}g((k+2)\beta_n)$ it follows from I (5.3) that for n sufficiently large,

$$f^{-1}(g((k+1)\beta_n) - g(k\beta_n)) \leq 2f^{-1}(g((k+2)\beta_n) - g((k+1)\beta_n))$$

from which follows (6.2.2) i). ∎

(6.4.9) Proposition

$L_g(\beta,1)$ is isomorphic to a subspace of $L_f(\alpha,0)$ iff

$$\exists\, \rho > 0 \text{ and } \tau \in (0,1) \ni \sup_n \frac{f(\rho\alpha_n)}{g(\tau\beta_n)} < \infty .$$

Proof

We argue exactly as in (6.4.8) except that $g(k\beta_n)$ is replaced by $g(\rho_k\beta_n)$ where $(\rho_k)$ is a sequence increasing to 1. ∎

(6.4.10) Proposition

$L_g(\beta,0)$ is isomorphic to a subspace of $L_f(\alpha,0)$ iff

i) $\exists\, M_1 > 0$ and a sequence $(\rho_k)$ decreasing to $0 \ni$

$$\overline{\lim_n} \frac{f^{-1}g(\rho_k\beta_n)}{f^{-1}g(\rho_{k+1}\beta_n)} \le M_1 \quad \forall k.$$

ii) $$\forall \tau > 0\ \exists\, \rho > 0 \ni \sup_n \frac{f(\rho\alpha_n)}{g(\tau\beta_n)} < \infty .$$

Proof

It is a straightforward computation to verify that ii) here is equivalent to (6.2.2) ii).

Suppose that i) holds. Then we have

$$\overline{\lim_n} \frac{f^{-1}g(\rho_k\beta_n)}{f^{-1}g(\rho_{k+1}\beta_n)} \le M_1 .$$

Then set $b_n^k = e^{-g(\rho_k\beta_n)}$ and let $M > M_1$. By I (5.3) we have, for n sufficiently large,

$$f^{-1}(\log\frac{b_n^{k+1}}{b_n^k}) \le f^{-1}g(\rho_k\beta_n) \le M_1 f^{-1}g(\rho_{k+1}\beta_n) \le Mf^{-1}(\tfrac{1}{2}g(\rho_{k+1}\beta_n))$$

$$\leq Mf^{-1}(g(\rho_{k+1}\beta_n) - g(\rho_{k+2}\beta_n)) = Mf^{-1}(\log\frac{b_n^{k+2}}{b_n^{k+1}})$$

and after the usual adjustment for small n we obtain (6.2.2) i).

Conversely if (6.2.2) ii) holds we set $c_n^k = e^{-g(\frac{1}{k}\beta_n)}$ and invoke (6.4.3) to obtain

$$\overline{\lim_n} \frac{f^{-1}(g(\frac{1}{j_k}\beta_n) - g(\frac{1}{j_{k+1}}\beta_n))}{f^{-1}(g(\frac{1}{j_{k+1}}\beta_n) - g(\frac{1}{j_{k+2}}\beta_n))} \leq M_1$$

and from I (5.3), it follows that

$$\overline{\lim_n} \frac{f^{-1}(g(\frac{1}{j_k}\beta_n))}{f^{-1}(g(\frac{1}{j_{k+1}}\beta_n))} \leq M_1 \quad .$$

and so we have i) with $\rho_k = \frac{1}{j_k}$ .

∎

(6.4.11) Proposition

$L_g(\beta,-1)$ is isomorphic to a subspace of $L_f(\alpha,0)$ iff

i) $\exists\ M_1 > 0$ and a sequence $(\rho_k)$ decreasing to 1 $\ni$

$$\overline{\lim_n} \frac{f^{-1}g(\rho_k\beta_n)}{f^{-1}g(\rho_{k+1}\beta_n)} \leq M_1 \quad \forall k$$

ii) $\forall \tau > 0\ \exists\ \rho > 0 \ni \sup_n \frac{f(\rho\alpha_n)}{g(\tau\beta_n)} < \infty$

Proof

It is a straightforward computation to verify that ii) here is equivalent to (6.2.2) ii).

Suppose that i) holds. Then we set $b_n^k = e^{-g(\rho_k \beta_n)}$ and let $M > M_1$. By I (5.3) we have, for n sufficiently large,

$$f^{-1}\left(\log\frac{b_n^{k+1}}{b_n^k}\right) \leq f^{-1}g(\rho_k\beta_n) \leq M_1 f^{-1}g(\rho_{k+1}\beta_n) \leq Mf^{-1}\left(\tfrac{1}{2}g(\rho_{k+1}\beta_n)\right)$$

$$\leq Mf^{-1}(g(\rho_{k+1}\beta_n) - g(\rho_{k+2}\beta_n)) = Mf^{-1}\left(\log\frac{b_n^{k+2}}{b_n^{k+1}}\right)$$

and after the usual adjustment we obtain (6.2.2) i).

Conversely, if (6.2.2) i) holds we set $c_n^k = e^{-g(\frac{k+1}{k}\beta_n)}$ and invoke (6.4.3) to obtain, with $\rho_k = \dfrac{j_k+1}{j_k}$,

$$\overline{\lim_n}\ \frac{f^{-1}(g(\rho_k\beta_n) - g(\rho_{k+1}\beta_n))}{f^{-1}(g(\rho_{k+1}\beta_n) - g(\rho_{k+2}\beta_n))} \leq M_1$$

and from I (5.3) it follows that

$$\overline{\lim_n}\ \frac{f^{-1}g(\rho_k\beta_n)}{f^{-1}g(\rho_{k+1}\beta_n)} \leq M_1 \quad .$$

∎

(6.5) We close with a brief mention of areas in which more work is necessary. In addition to the cases $r = \pm 1$ mentioned above, it seems that there should be results similar to those in section (2.3). For example, we can describe (2.3.4) roughly by saying that the finite type power series spaces are the "smallest" spaces with bases of type $(d_2)$. Do we have similar results for $L_f(\alpha,r)$ spaces?

In addition to generating results of this kind, it is also an interesting problem as to how to organize them. There should be a pattern which, in turn, suggests other results.

## 7. Notes and Remarks

There is nothing in this chapter about subspaces of a space which is a countable product of copies of a Köthe space. Thus is omitted the fundamental result of Kōmura and Kōmura [41] and some generalizations of it due to Ramanujan and Terzioglu [54]. Also there are some interesting connections with this point in the work of Vogt and Wagner [61] and [63]. More generally, one might consider a countable product of (perhaps different) Köthe spaces, or what is the same thing a nuclear Fréchet space with a basis but not a continuous norm.

A more important restriction in this chapter is that we only analyse the subspaces with bases of a nuclear Fréchet space. An entirely different and very important approach is found in the work of Vogt and Wagner [60], [61], and [63] who do not assume the existence of a basis. Thus they obtain the results of (2.2) in a more general context. Their methods have not been applied to the other cases studied in this chapter and it will be interesting to see what will be the results, especially with regard to Sections 3, 4 and 5.

The fundamental inequality is implicit in the work of the author [29], [31] and Alpseymen [1]. Essentially our approach to the subspace question can be roughly described as a synthesis of this condition (necessary) and the condition of II (2.2.3). The basic idea of our construction is to decompose a stable space into infinitely many infinite stepspaces, each of which is then weakly stable. Then in each stepspace we use (1.3) and II (2.2.3) to determine necessary and sufficient conditions for embedding a one-dimensional subspace in this stepspace. Of course this must be done uniformly as the stepspace varies and it is here that the nuclearity comes in, for example, in condition ii) of (2.2.1) or (2.2.2).

Theorem (2.2.1) is due to Alpseymen [1] and theorem (2.2.2) was proved in [31]. The earliest results in this direction appeared more or less simultaneously in [29] and [60] although in spirit the ideas began to germinate in [26], [27] and [28]. In a certain sense much of the approach described in these lectures was inspired by a reading of Bessaga's proof of Dragilev's theorem [4], especially lemma 2.0.

The results of section (2.4) are a culmination of the investigations in [26] and [27]. An interesting result in this direction is due to Kaširin [39]. The results of section (2.5) appear here for the first time.

The results of Section 3 are a somewhat streamlined version of material in [27] except that there, only subspaces of infinite type power series spaces are considered. In our context, the same arguments work for both types. Thus, some of the results in this section are, technically, new. We do not mention the analogue of (3.3), that is, no finite type power series space is a subspace of an infinite type power series space. This was proved by Zahariuta [65] using much simpler methods.

The results of Section 4 appear here for the first time. There are no results in Section 5.

All of the results of Section 6 are due to the author jointly with Alpseymen and De-Grande De-Kimpe [2]. In sections (6.3) and (6.4) we have presented all 12 cases in the interest of completeness and symmetry. Actually, 4 of these cases, (6.3.1), (6.3.4), (6.4.6) and (6.4.7) were first proved by Zahariuta [65] again using simpler methods.

# CHAPTER IV

## QUOTIENT SPACES

### 1. The Fundamental Inequality

(1.1) Our main interest in this chapter is to determine conditions which characterize those Köthe spaces which are isomorphic to a quotient space of a given Köthe space. Our knowledge is, at the moment, rather less than in the case of subspaces. Also, although the general arguments are, essentially, dual to the arguments for subspaces, there are several cases in which serious new technical difficulties arise, and we have not removed them all as yet.

The results are strikingly analogous to the subspace case. Again we have, basically, two conditions which are very closely related to the ones we discussed in Chapter III.

The tools developed in Chapter II are for constructions so they represent sufficient conditions. For the necessity we have, of course, I (6.2.4) and an inequality which we now derive.

(1.2) The context and the argument is slightly different than in III (1.2), (1.3). Let E, F be nuclear Fréchet spaces with bases $(x_n)$, $(y_n)$ respectively and let T: E → F be a continuous linear surjection. Let a be a representation of $(x_n)$ and define

$$p_n^k = \min\{p: \inf_i \frac{a_i^k}{|t_i^n|} = \frac{a_p^k}{|t_p^n|}\}$$

and $p_n^k = \infty$ if this set is empty. Let the matrix $(t_i^n)$ be given by the relation

$$T(x_i) = \sum_n t_i^n y_n, \quad i \in \mathbb{N} .$$

(1.3) Lemma

If F admits a continuous norm then there exists $k_0 \ni p_n^k < \infty \ \forall\, n$ and $k \geq k_0$. Moreover, if we define $b = (b_n^k)$ by

$$b_n^k = \frac{a_k^k / p_n}{|t_{p_n^k}^n|}, \quad k \geq k_0$$

then b is a representation of $(y_n)$.

Proof

Let $||\cdot||$ be a continuous norm on F. From the continuity of T and the nuclearity of F we have a norm $|\cdot|$ on F, $C > 0$ and $k_0 \ni$

$$|t_i^m| \, ||y_m|| \leq \sum_n |t_i^n| \, ||y_n|| \leq |\sum_i t_i^n y_n| = |Tx_i| \leq Ca_i^{k_0} \quad \forall\ i,m$$

from which it follows that $p_n^k < \infty \ \forall\ n$ and $k \geq k_0$.

For the second part we observe that since T is onto then a subset of F' is bounded iff its image under the transpose T' is bounded in E'. Thus if $(A_k)$ is a fundamental system of closed, absolutely convex, bounded sets in E', and $B_k = (T')^{-1}(A_k)$ then $(B_k)$ is a fundamental system of bounded sets in F' so $(B_k^0)$ is a fundamental system of nbds of 0 in F. We take $A_k = U_k^0$ where

$$U_k = \{x = \sum_i \xi_i x_i \in E: \sum_i |\xi_i| a_i^k \leq 1\} .$$

Let $(g_n)$ be the sequence of functionals biorthogonal to $(y_n)$. Since F admits a continuous norm $(g_n)$ is bounded say $g_n \in B_k \ \forall\ n$ and $k \geq k_0$ (increasing $k_0$ if necessary). Then using $\mu_S$ to denote the gauge of S we have, for $k \geq k_0$,

$$\mu_{A_k}(T'(q_n)) = \inf\{\lambda > 0: T'(q_n) \in \lambda A_k\}$$

$$= \inf\{\lambda > 0: |t_i^n| \leq \lambda a_i^k \ \forall i\} = \sup_i \frac{|t_i^n|}{a_i^k} = \frac{1}{b_n^k} .$$

Clearly $\mu_{A_k}(T'g_n) = \mu_{B_k}(g_n)$ and $1 = |g_n(y_n)| \leq \mu_{B_k}(g_n)\mu_{B_k^0}(y_n)$ so

$$b_n^k \leq \mu_{B_k^0}(y_n) \quad .$$

On the other hand, since $(y_n)$ is a basis for the Fréchet space F, $(g_n(y)y_n)_n$ is bounded for each $y \in F$ so by the uniform boundedness principle, given k there exists j and $M > 0 \ni$

$$|g_n(y)|\mu_{B_k^0}(y_n) \leq M\mu_{B_j^0}(y) \qquad \forall n \text{ and } y \in F,$$

which implies that $\mu_{B_k^0}(y_n)g_n \in MB_j^{00} = MB_j$ so that

$$\mu_{B_k^0}(y_n) \leq \frac{M}{\mu_{B_j}(g_n)} = M\, b_n^j$$

and since $(\mu_{B_k^0}(y_n))$ is a representation of $(y_n)$ we are finished. ∎

(1.4) Lemma

If F admits a continuous norm then in the context of (1.2), (1.3) we have

$$\frac{a^j_{p_n^j}}{a^k_{p_n^j}} \leq \frac{b_n^j}{b_n^k} \leq \frac{a^j_{p_n^k}}{a^k_{p_n^k}} \qquad \forall n,\ k,\ j\ .$$

<u>Proof</u>

We have

$$\frac{b_n^j}{b_n^k} = \frac{a^j_{p_n^k}\,|t^n_{p_n^k}|}{|t^n_{p_n^j}|\,a^k_{p_n^k}}$$

and by definition of $p_n^k$ this quantity is increased if $p_n^j$ is replaced by $p_n^k$ and decreased if $p_n^k$ is replaced by $p_n^j$.

∎

## 2. Stable Power Series Spaces

(2.1) We assume (see III (2.1)) that $\sup\limits_n \frac{\alpha_{2n}}{\alpha_n} = A < \infty$ so $\alpha$ is stable. This will do for our preliminary considerations but in the case of infinite type power series spaces we will have to make additional assumptions on $\alpha$ to obtain the characterization. For finite type, there is a gap between our necessary and our sufficient conditions which cannot be resolved by further restrictions on $\alpha$.

(2.2) Characterizations

(2.2.1) Lemma

Let $\rho_k > 0$, $x_n^k > 0 \;\forall\; k, n$ and suppose that $\forall\; k$ there exists $n_k$ and a bijection $\pi^k\colon \mathbb{N} \to \mathbb{N} \ni$

$$x^k_{\pi^k(n)} \leq e^{-\rho_k \alpha_n}, \quad n \geq n_k .$$

Then there exists a single bijection $\pi\colon \mathbb{N} \to \mathbb{N} \ni$

$$x^k_{\pi(n)} \leq e^{-A^{-k}\rho_k \alpha_n}, \quad n \geq 2^k n_k .$$

Proof

Let $\beta\colon \mathbb{N} \to \mathbb{N}\times\mathbb{N}$ be the bijection given by

$$\beta(n) = (\beta_1(n), \beta_2(n)) = (k,m) \text{ where } n = 2^{k-1}(2m-1).$$

Let $\tilde{\pi}\colon \mathbb{N} \to \mathbb{N}$ be the surjection given by $\tilde{\pi}(n) = \pi^{\beta_1(n)}(\beta_2(n))$ and let $\pi\colon \mathbb{N} \to \mathbb{N}$ be the bijection obtained from $\tilde{\pi}$ by looking at each $\tilde{\pi}(n)$ successively for $n = 1,2,\ldots,$ and deleting each repetition. Now for each $k,m$ $\pi^k(m) = \pi(n)$ for a certain $n$. Indeed, we obtain $\pi^k(m) = \pi(n)$ either when $n = 2^{k-1}(2m-1)$ or sooner (a smaller value of $n$) if this is a repetition. Hence,

$$\pi(n) = \pi^k(m) \text{ with } n \leq 2^{k-1}(2m-1) \quad \forall\, k,\, m.$$

Hence, given k, n there exists $m \leq 2^{k-1}(2m-1) \leq 2^k m \ni$

$$x^k_{\pi(n)} = x^k_{\pi^k(m)} \leq e^{-\rho_k \alpha_m} \leq e^{-A^{-k}\rho_k \alpha_n} \quad m \geq n_k .$$

Since $n \geq 2^k n_k$ implies $m \geq n_k$ we are finished. ∎

(2.2.2) Lemma

Let F be a nuclear Fréchet space with a type $(d_4)$ basis $(y_n)$ and assume that ∀ nbds of 0 U in F and $\rho > 0$ there exists nbd of 0 V in F ∋

$$d_n(V,U) \leq e^{-\rho\alpha_n} \quad n \text{ sufficiently large} .$$

Then ∀ C,D > 0 there exists a permutation and diagonal transform of $(y_n)$ which has a representation b ∋ there exists a sequence $(k_n)$ with $\lim_n k_n = \infty$ and

$$Ce^{\alpha_n} \leq \frac{b_n^{k+2}}{b_n^{k+1}} \leq \frac{1}{D}\frac{b_n^{k+1}}{b_n^k}, \quad k \leq k_n .$$

Proof

Let a be a $(d_4)$ representation of $(y_n)$. By our assumption and the computation of Kolmogorov diameters it follows that ∀ k there exists j and a bijection $\pi^k: \mathbb{N} \to \mathbb{N} \ni$

$$e^{2kA^k\alpha_n} \leq \frac{a^k_{\pi^k(n)}}{a^k_{\pi^k(n)}} \quad n \text{ sufficiently large.}$$

Applying (2.2.1) we have a single permutation $\pi$ which we can eliminate by considering that a is a representation of a permutation of $(y_n)$ so we have,

$$\forall\, k \;\exists\, j \ni e^{2k\alpha_n} \le \frac{a_n^j}{a_n^k} \quad n \text{ sufficiently large} \tag{1}$$

We decompose $\mathbb{N}$ into a union of pairwise disjoint sets $\mathbb{N}_\nu$, $0 \le \nu \le \infty$ by setting

$$\mathbb{N}_0 = \{n: \frac{a_n^2}{a_n^1} < e^{2\alpha_n}\}$$

$$\mathbb{N}_\nu = \{n: \frac{a_n^{\nu+2}}{a_n^{\nu+1}} < e^{2\alpha_n} \le \frac{a_n^{\nu+1}}{a_n^{\nu}}\} \qquad 1 \le \nu < \infty$$

$$\mathbb{N}_\infty = \{n: e^{2\alpha_n} \le \frac{a_n^{k+1}}{a_n^k} \quad \forall k\}\ .$$

We can replace $(y_n)$ by a diagonal transform so that $a_n^{\nu+1} = 1$ if $0 \le \nu < \infty$ and $n \in \mathbb{N}_\nu$. We define b by

$$b_n^k = \begin{cases} e^{2(k-\nu-1)\alpha_n} & \text{if } 0 \le \nu < \infty,\ n \in \mathbb{N}_\nu \text{ and } k > \nu+1 \\ a_n^k & \text{otherwise} \end{cases}$$

Now if $0 \le \nu < \infty$, $n \in \mathbb{N}_\nu$ and $k > \nu+1$ then by the $(d_4)$ condition,

$$a_n^k = \frac{a_n^k}{a_n^{k-1}} \cdots \frac{a_n^{\nu+2}}{a_n^{\nu+1}} \le \left(\frac{a_n^{\nu+2}}{a_n^{\nu+1}}\right)^{k-\nu-1} < e^{2(k-\nu-1)\alpha_n} = b_n^k$$

and, taking j from (1), we have for n sufficiently large,

$$b_n^k = e^{2(k-\nu-1)\alpha_n} \le e^{2k\alpha_n} \le \frac{a_n^j}{a_n^k} \le \frac{a_n^j}{a_n^{\nu+1}} = a_n^j$$

so b is a representation of a permutation and diagonal transform of $(y_n)$. We will show that

$$e^{2\alpha_n} \leq \frac{b_n^{k+2}}{b_n^{k+1}} \leq \frac{b_n^{k+1}}{b_n^{k}} \qquad \forall\ n,\ k \tag{2}$$

From the definition of b and the $(d_4)$ property of a, the second inequality is immediate for all k except $k = \nu, \nu+1$, $n \in \mathbb{N}_\nu$, $0 \leq \nu < \infty$.

If $k = \nu$,

$$\frac{b_n^{k+2}}{b_n^{k+1}} = \frac{e^{2\alpha_n}}{1} \leq \frac{a_n^{\nu+1}}{a_n^{\nu}} = \frac{b_n^{k+1}}{b_n^{k}}$$

and if $k = \nu + 1$,

$$\frac{b_n^{k+2}}{b_n^{k+1}} = e^{2\alpha_n} = \frac{b_n^{k+1}}{b_n^{k}}$$

which establishes the second inequality in (2). It also says that it suffices to prove the first inequality for large k where it is an equality.

Finally we get the desired result if we replace $b_n^k$ by $\frac{1}{D^{1+2+...+k}} b_n^k$, and use the fact that $\lim_n \alpha_n = \infty$. ∎

(2.2.3) We will say that α is <u>multiplicatively stable</u> if

$$\sup_n \frac{\alpha_{n^2}}{\alpha_n} < \infty \quad .$$

It is easy to see that this is equivalent to the condition that $\Lambda_\tau(\alpha) \cong \Lambda_\tau(\alpha) \otimes \Lambda_\tau(\alpha)$, $\tau = 1,\infty$. We shall say that α is <u>sublinear</u> if

$$\sup_n (\alpha_{n+1} - \alpha_n) = B < \infty \ .$$

For example, if $\alpha_n = \log n$ or $\alpha_n = (\log n)^2$ then α is both multiplicatively stable and sublinear. But if $\alpha_n = n$ then α is sublinear but not multiplicatively stable and if

$$\alpha_n = \begin{cases} 2 \log n & \text{if } n \text{ is a perfect square} \\ \log n & \text{otherwise} \end{cases}$$

then $\alpha$ is multiplicatively stable but not sublinear.

(2.2.4) Theorem

Assume that $\alpha$ is multiplicatively stable and sublinear. Let F be a nuclear Fréchet space with basis $(y_n)$, and a continuous norm.

Then F is isomorphic to a quotient space of $\Lambda_\infty(\alpha)$ iff

i) The basis $(y_n)$ is of type $(d_4)$

ii) $\forall$ nbds of 0 U in F and $\rho > 0$ there exists nbd of 0 V in F $\ni$

$$d_n(V,U) \leq e^{-\rho\alpha_n} \quad \text{for n sufficiently large}$$

Proof

The necessity of ii) follows from I (6.2.4) and the fact that it holds for $\Lambda_\infty(\alpha)$. For i) we apply (1.4) with $a_n^k = e^{k\alpha_n}$. This yields a representation b for $(y_n) \ni$

$$\frac{b_n^{k+2}}{b_n^{k+1}} \leq e^{\alpha_{p_n^{k+1}}} \leq \frac{b_n^{k+1}}{b_n^k} \qquad \forall n, k$$

which is the $(d_4)$ condition.

Now we assume that i), ii) hold and apply (2.2.2) with $C = e^{\alpha_1}$ and $D = e^B$. Thus we have $(k_n)$.

Since $\alpha$ is multiplicatively stable it suffices to construct a quotient space of $\Lambda_\infty(\alpha)\otimes\Lambda_\infty(\alpha)$ isomorphic to K(b). We may consider the basis $(e_m \otimes e_\nu)_{m,\nu}$ in $\Lambda_\infty(\alpha)\otimes\Lambda_\infty(\alpha)$ which has the representation $(e^{k(\alpha_m+\alpha_\nu)})_{k,(m,\nu)}$. Let us fix $m \in \mathbb{N}$ and write $a_\nu^k = e^{k(\alpha_m+\alpha_\nu)}$. We will select a strictly decreasing set of indices $p^k$, $1 \leq k \leq k_m$ and a corresponding set of scalars $t_{p^k}$, $1 \leq k \leq k_m$.

First we show that we can select $p^1 > p^2 > p^{k_m} \ni$

$$e^{\alpha_m+\alpha_{p^{k+1}}} = \frac{a^{k+1}_{p^{k+1}}}{a^{k}_{p^{k+1}}} \le \frac{b^{k+1}_m}{b^{k}_m} < \frac{a^{k+1}_{p^k}}{a^{k}_{p^k}} = e^{\alpha_m+\alpha_{p^k}} \qquad k \le k_m.$$

We begin by choosing $p^1$ so that the right hand inequality holds. Suppose that we have chosen $p^1,\ldots,p^k$, $k < k_m$ so that the right inequality and also the left inequality holds. Then we let $p^{k+1}$ be the largest index $\ni$ the left inequality holds. To see that this is possible we must show that it holds if $p^{k+1} = 1$. For this we have, from (2.2.2)

$$\frac{a^{k+1}_1}{a^k_1} = e^{\alpha_1}e^{\alpha_m} = Ce^{\alpha_m} \le \frac{b^{k+1}_m}{b^k_m} .$$

Then we must check the right inequality for $p^{k+1}$.

$$\frac{b^{k+2}_m}{b^{k+1}_m} \le \frac{1}{D}\frac{b^{k+1}_m}{b^k_m} < e^{-B+\alpha_m+\alpha_{p^{k+1}+1}} \le e^{\alpha_m+\alpha_{p^{k+1}}} = \frac{a^{k+2}_{p^{k+1}}}{a^{k+1}_{p^{k+1}}} .$$

Finally we observe that since the left inequality does not hold with $p^k$ it follows that $p^{k+1} < p^k$.

This completes the selection of $p^1,\ldots,p^{k_m}$ and we define

$$t_{p^k} = \frac{a^k_{p^k}}{b^k_m}, \quad k \le k_m$$

from which it follows immediately that

$$\frac{a^{k+1}_{p^{k+1}}}{a^{k+1}_{p^k}} < \frac{|t_{p^{k+1}}|}{|t_{p^k}|} \le \frac{a^{k}_{p^{k+1}}}{a^{k}_{p^k}}$$

so by II (2.3.3) it follows that $p^k(t_{p^1},\dots,t_{p^{k_m}}) = p^k$, $1\leq k\leq k_m$.

Hence by II (3.3.1) there is a quotient map T: $\Lambda_\infty(\alpha)\otimes\Lambda_\infty(\alpha) \to K(b)$. ∎

(2.2.5) Theorem

Assume that $\alpha$ is multiplicatively stable and sublinear. Let E be a nuclear Fréchet space with basis $(y_n)$. Then E is isomorphic to a quotient of $\Lambda_\infty(\alpha)$ iff E is isomorphic to a space of one of the following kinds:

i) F where F has continuous norm and satisfies i), ii) of (2.2.4)

ii) $\omega$

iii) $F\times\omega$ where F is as in i)

iv) $\prod_{\nu=1}^{\infty} F_\nu$ where each $F_\nu$ is as in i)

Proof

Suppose that E is a quotient of $\Lambda_\infty(\alpha)$ and let a be a representation of $(y_n)$ set $\mathbb{N}_0 = \phi$ and $\mathbb{N}_\nu = \{n \in \mathbb{N}\sim\mathbb{N}_{\nu-1}: a_n^k>0\}$, $\nu = 1,2,\dots$, and let $\sigma$ be the set of $\nu \ni \mathbb{N}_\nu \neq \phi$. Clearly, $\mathbb{N} = \bigcup_{\nu\in\sigma} \mathbb{N}_\nu$.

Let $a(\nu) = a|_{\mathbb{N}_\nu}$, $\nu\in\sigma$ so $K(a(\nu))$ has a continuous norm and is a quotient of $\Lambda_\infty(\alpha)$ so it satisfies i), ii) of (2.2.4). Consider the map T: $K(a) \to \prod_{\nu\in\sigma} K(a(\nu))$ by $T(\xi) = (\xi|_{\mathbb{N}_\nu})_{\nu\in\sigma}$. To see that this map is an isomorphism it suffices to show that it is onto. Indeed, if $\xi(\nu) \in K(a(\nu))$, $\nu\in\sigma$ let $\xi = (\xi_n)_{n\in\mathbb{N}}$ where $(\xi_n)_{n\in\mathbb{N}_\nu} = \xi(\nu)$. Then, given any k, $a_n^k = 0$ if $\nu > k$ so

$$\sum_n |\xi_n| a_n^k = \sum_{\substack{\nu\in\sigma\\ \nu\leq k}} \sum_{n\in\mathbb{N}_\nu} |\xi_n| a_n^k < \infty.$$

Finally, referring to our four conclusions, we get i) if $\sigma$ is finite, ii) if each $\mathbb{N}_\nu$ is finite, iii) if finitely many $\mathbb{N}_\nu$ are infinite and iv) otherwise.

Conversely, suppose these four conditions hold. If i) holds then we can apply (2.2.4). For ii) we must show that $\omega$ is a quotient of $\Lambda_\infty(\alpha)$. This is a standard result which can be shown in many ways. For example, a proof is sketched in [57, p. 192]. Alternatively, we could prove an "infinite" version of II (3.3.1) in which $(t_{p_{n-1}+1},\ldots,t_{p_n})$ is replaced by an infinite set chosen so that $b_n^k = 0$ for $k < n$.

If iii) holds we use the above discussion and the fact that $\Lambda_\infty(\alpha) \cong \Lambda_\infty(\alpha) \times \Lambda_\infty(\alpha)$.

Finally if iv) holds we apply (2.2.4) for each $\nu$ to obtain quotient maps $\Lambda_\infty(\alpha) \to F_\nu$ so we have a single quotient map $\omega \otimes \Lambda_\infty(\alpha) = \prod_{\nu=1}^{\infty} \Lambda_\infty(\alpha) \to \prod_{\nu=1}^{\infty} F_\nu$. We then take the tensor product of the identity $\Lambda_\infty(\alpha) \to \Lambda_\infty(\alpha)$ and the above quotient map $\Lambda_\infty(\alpha) \to \omega$. By a result of Grothendieck [37] this gives a quotient map $\Lambda_\infty(\alpha) \cong \Lambda_\infty(\alpha) \otimes \Lambda_\infty(\alpha) \to \omega \otimes \Lambda_\infty(\alpha)$. We finish by composing the two quotient maps. ∎

(2.2.6) Turning to quotients of finite type power series spaces, our situation deteriorates in that we no longer get a characterization. We are able to extablish the expected necessary condition but we are unable to make the embedding without strengthening the condition. One compensation is that our results are still good enough to determine all power series and all $L_f(\alpha,r)$ quotient spaces of a stable finite type power series space. Another compensation is that we get by with stability and do not have to assume anything stronger.

Hopefully, additional research will eliminate the difficulties in this section. Because this part is especially tentative we will not try for the most general results but restrict ourselves to spaces with continuous norm.

(2.2.7) Proposition

Let F be a nuclear Fréchet space with a basis $(y_n)$ and a continuous norm and assume that F is isomorphic to a quotient space of $\Lambda_1(\alpha)$. Then the basis $(y_n)$ is type $(d_6)$ and $\forall$ nbds of 0 U in F there exists nbd of 0 V in F and $\rho > 0$ $\ni$

$$d_n(V,U) \leq e^{-\rho\alpha_n}, \quad \text{for n sufficiently large}$$

Proof

The statement about the diameters follows as usual from I (6.2.4), (6.3.2) and the fact that it holds for $\Lambda_1(\alpha)$. To see that the basis is type $(d_6)$ let $M \geq 1$ and apply (1.4) with $a_n^k = e^{-\frac{1}{\rho_k}\alpha_n}$ where $(\rho_k)$ is a strictly increasing unbounded sequence of positive numbers $\ni$ $(1+M)\rho_k \leq \rho_{k+1}$. Then, for any $\ell$, k

$$M(\frac{1}{\rho_{k+1}} - \frac{1}{\rho_\ell}) \leq \frac{M}{\rho_{k+1}} \leq \frac{1}{\rho_k} - \frac{1}{\rho_{k+1}},$$

so (1.4) gives a representation b for $(y_n)$ $\ni$

$$\left(\frac{b_n^\ell}{b_n^{k+1}}\right)^M \leq e^{M(\frac{1}{\rho_{k+1}} - \frac{1}{\rho_\ell})\alpha} p_n^{k+1} \leq e^{(\frac{1}{\rho_k} - \frac{1}{\rho_{k+1}})\alpha} p_n^{k+1} \leq \frac{b_n^{k+1}}{b_n^k} \quad \forall n,k,\ell.$$

∎

(2.2.8) Proposition

Let F be a nuclear Fréchet space with a continuous norm and a basis $(y_n)$ of type $(d_6)$ and assume that $\forall$ nbds of 0 U in F and $\rho > 0$ there exists nbd of 0 V in F $\ni$

$$d_n(V,U) \leq e^{-\rho\alpha_n}, \quad \text{for n sufficiently large.}$$

Then F is isomorphic to a quotient space of $\Lambda_1(\alpha)$.

Proof

Let c be a representation of $(y_n) \ni$ the $(d_6)_M$ condition holds with M = 2A. We apply our condition on the Kolmogorov diameters and (2.2.1) so we have a permutation $\pi$, a subsequence of indices $(j_k)$ and indices $(n_k) \ni$

$$e^{\alpha_n} \leq \frac{c^{j_{k+1}}_{\pi(n)}}{c^{j_k}_{\pi(n)}} \qquad \forall\ n \geq n_k.$$

Applying II (1.5.2) we have a representation b for a permutation of $(y_n)$ and a sequence of indices $(k_n) \ni \lim_n k_n = \infty$ and

$$e^{\alpha_n} \leq \frac{b^{k+2}_n}{b^{k+1}_n} \leq \left(\frac{b^{k+1}_n}{b^k_n}\right)^{1/M}, \qquad k \leq k_n.$$

Now let m be fixed and set $\beta_\nu = \alpha_{2^{\nu-1}(2m-1)}$, $a^k_\nu = e^{-\frac{1}{\rho_k}\beta_\nu}$ where $\rho_k = A2^{k-1}$. We will select $p^1 > p^2 > \ldots > p^{k_m} \ni$

$$e^{\left(\frac{1}{\rho_k} - \frac{1}{\rho_{k+1}}\right)\beta_{p^{k+1}}} = \frac{a^{k+1}_{p^{k+1}}}{a^k_{p^{k+1}}} \leq \frac{b^{k+1}_m}{b^k_m} < \frac{a^{k+1}_{p^k}}{a^k_{p^k}} = e^{\left(\frac{1}{\rho_k} - \frac{1}{\rho_{k+1}}\right)\beta_{p^k}}, \quad k \leq k_m.$$

First we select $p_1 \ni$ the right inequality holds. Suppose that we have chosen $p_1, \ldots, p_k$, $k < k_m \ni$ the right inequality and also the left inequality holds. Then we let $p_{k+1}$ be the largest index $\ni$ the left inequality holds. To see that this is possible we must show that it holds if $p^{k+1} = 1$. We have, for $k \leq k_m$,

$$e^{\left(\frac{1}{\rho_k} - \frac{1}{\rho_{k+1}}\right)\beta_1} \leq e^{\frac{1}{A}\alpha_{2m-1}} \leq e^{\alpha_m} \leq \frac{b^{k+1}_m}{b^k_m}$$

Then we must check the right inequality with k replaced by k + 1.

$$\frac{b_m^{k+2}}{b_m^{k+1}} \le \left(\frac{b_m^{k+1}}{b_m^{k}}\right)^{\frac{1}{M}} < e^{(\frac{1}{\rho_k} - \frac{1}{\rho_{k+1}})\frac{1}{M}\beta p^{k+1}+1} \le e^{\frac{1}{2^{k+1}}A\beta p^{k+1}} =$$

$$e^{(\frac{1}{\rho_{k+1}} - \frac{1}{\rho_{k+2}})\beta p^{k+1}} .$$

Finally we observe that since the left inequality does not hold with $p^k$ it follows that $p^{k+1} < p^k$.

This completes the selection of $p^1,\dots,p^{k_m}$ and we define

$$t_{p^k} = \frac{a^k_{p^k}}{b^k_m},\ k \le k_m$$

from which it follows immediately that

$$\frac{a^{k+1}_{p^{k+1}}}{a^{k+1}_{p^k}} < \frac{|t_{p^{k+1}}|}{|t_{p^k}|} \le \frac{a^k_{p^{k+1}}}{a^k_{p^k}} .$$

so by II (2.3.3) it follows that $p^k(t_{p^1},\dots,t_{p^{k_m}}) = p^k$, $1 \le k \le k_m$. Hence by II (3.3.1), there is a quotient map $T\colon \Lambda_1(\alpha) \to K(b)$.

∎

(2.3) Some consequences

(2.3.1) The main consequence that we shall obtain from the results of the previous section concern a property to which we shall return in Chapter V.

In this section we make no assumptions about $\alpha$ being stable but only assume that it is a nuclear exponent sequence of appropriate type. This is because everything that we will be using is a consequence of (1.4) or II (1.3) where no stability assumptions were required.

(2.3.2) Proposition

Let F be a nuclear Fréchet space with a basis $(y_n)$ which is isomorphic to a subspace of an infinite type power series space and to a quotient space of an infinite type power series space. Then F is isomorphic to an infinite type power series space.

Proof

By III (2.2.1), $(y_n)$ is type $(d_3)$. Being a subspace of a space with a continuous norm, F admits a continuous norm. By (2.2.4), $(y_n)$ is type $(d_4)$ so the result follows from II (1.4.2).

∎

(2.3.3) Proposition

Let F be a nuclear Fréchet space with a basis $(y_n)$ which is isomorphic to a subspace of a finite type power series space and to a quotient space of a finite type power series space. Then F is isomorphic to a finite type power series space.

Proof

By III (2.2.2), $(y_n)$ is type $(d_5)$. Being a subspace of a space with a continuous norm, F admits a continuous norm. By (2.2.7), $(y_n)$ is type $(d_6)$ so the result follows from II (1.4.5).

∎

(2.4) Power series quotient spaces of power series spaces

(2.4.1) Corollary

Assume that $\alpha$ is multiplicatively stable and sublinear. Then $\Lambda_\infty(\beta)$ is isomorphic to a quotient space of $\Lambda_\infty(\alpha)$ iff $\sup_n \frac{\alpha_n}{\beta_n} < \infty$.

Proof

The coordinate basis in $\Lambda_\infty(\beta)$ is $(d_4)$ and the condition here is equivalent to ii) of (2.2.4).

∎

(2.4.2) Corollary

Assume that $\alpha$ is multiplicatively stable and sublinear. Then $\Lambda_1(\beta)$ is isomorphic to a quotient space of $\Lambda_\infty(\alpha)$ iff $\lim_n \frac{\alpha_n}{\beta_n} = 0$.

Proof

The coordinate basis in $\Lambda_1(\beta)$ is $(d_4)$ and ii) of (2.2.4) becomes,

$$\forall k,\ \rho > 0\ \exists\ j \ni \rho\alpha_n \leq \left(\frac{1}{k} - \frac{1}{j}\right)\beta_n \quad n \text{ sufficiently large}$$

which is clearly equivalent to the given condition. ∎

(2.4.3) Corollary

No infinite type power series space is isomorphic to a quotient of a finite type power series space.

Proof

If it were, then by (2.2.7) the coordinate basis would be type $(d_6)$ so by II (1.4.4) it would be type $(d_2)$ but it is type $(d_1)$ which is impossible. ∎

(2.4.4) Corollary

Assume that $\alpha$ is stable. Then $\Lambda_1(\beta)$ is isomorphic to a quotient space of $\Lambda_1(\alpha)$ iff $\sup\limits_n \frac{\alpha_n}{\beta_n} < \infty$.

Proof

If $\Lambda_1(\beta)$ is a quotient of $\Lambda_1(\alpha)$ the conclusion follows from (2.2.7). On the other hand, if $\sup\limits_n \frac{\alpha_n}{\beta_n} < \infty$ then by I (4.5), $\Lambda_1(\beta) \cong \Lambda_1((\alpha_{j_n}))$, $j_n < j_{n+1}$ so this latter space is a complemented subspace and hence a quotient space of $\Lambda_1(\alpha)$. ∎

(2.5) $L_f(\beta,r)$ quotient spaces of power series spaces

(2.5.1) Because of considerable variation we make no blanket stability assumptions on $\alpha$, except of course that it be nuclear of the appropriate type.

(2.5.2) Corollary

If $0 < r \leq \infty$ then $L_f(\beta,r)$ is not isomorphic to a quotient space of $\Lambda_\infty(\alpha)$.

Proof

In (2.2.4) we do not use the stability of $\alpha$ for necessity so if $L_f(\beta,r)$ is isomorphic to a quotient of $\Lambda_\infty(\alpha)$ then the coordinate basis in $L_f(\beta,r)$ would be type $(d_4)$. We know that this space is type $(d_1)$ so by II (1.4.1), (1.4.2) $L_f(\beta,r)$ would be a power series space but it is not. ∎

(2.5.3) Corollary

Let $\alpha$ be multiplicatively stable and sublinear and let $-\infty < r \leq 0$. Then $L_f(\beta,r)$ is isomorphic to a quotient space of $\Lambda_\infty(\alpha)$ iff

$$\sup_n \frac{\alpha_n}{f(\rho_0\beta_n)} < \infty \qquad \forall \rho_0 > -r .$$

Proof

The space $L_f(\beta,r) \cong K(b)$ where $b_n^k = e^{-f(r_k\beta_n)}$ and $(r_k)$ is a strictly decreasing sequence which converges to $-r$. The $(d_4)$ condition is

$$f(r_k\beta_n) - f(r_{k+1}\beta_n) \geq f(r_{k+1}\beta_n) - f(r_{k+2}\beta_n) \quad \forall n,k.$$

This is clear for n sufficiently large so we can adjust b so that it holds $\forall n$. This is condition i) of (2.2.4).

Condition ii) of (2.2.4) is, using b, equivalent to

$$\forall \rho_1 > -r,\ \rho > 0\ \exists\ \rho_2 > -r \ni \rho^{\alpha_n} \leq f(\rho_1\beta_n) - f(\rho_2\beta_n), \text{ n sufficiently large}$$

This clearly implies our condition. If, on the other hand, we have $\rho_0 > -r$ and $B \ni$

$$\alpha_n \leq Bf(\rho_0\beta_n) \quad \forall n$$

then, given $\rho_1 > -r$, $\rho > 0$ choose $\rho_2, \rho_0 \ni -r < \rho_2, \rho_0 < \rho_1$. By the fact that f is rapidly increasing, then for n sufficiently large,

$$f(\rho_1\beta_n) - f(\rho_2\beta_n) \geq B\rho f(\rho_0\alpha_n) \geq \rho\alpha_n$$

which establishes ii) of (2.2.4). ∎

(2.5.4) Corollary

If $0 < r \leq \infty$ then $L_f(\beta,r)$ is not isomorphic to a quotient space of $\Lambda_1(\alpha)$.

Proof

If this were so then by (2.2.7), the coordinate basis in $L_f(\beta,r)$ would be of type $(d_6)$ and so, by II (1.4.4) it would be type $(d_2)$. But we know this basis is type $(d_1)$ which is a contradiction. ∎

(2.5.5) Corollary

Let $\alpha$ be stable and $-\infty < r \leq 0$. Then $L_f(\beta,r)$ is isomorphic to a quotient space of $\Lambda_1(\alpha)$ iff

$$\sup_n \frac{\alpha_n}{f(\rho_0\beta_n)} < \infty \quad \forall \rho_0 > -r$$

Proof

The space $L_f(\beta,r) \cong K(b)$ where $b_n^k = e^{-f(r_k\beta_n)}$ and $(r_k)$ is a strictly decreasing sequence which converges to $-r$. As in (2.5.3) this gives the $(d_6)$ condition. The remaining condition in (2.2.8) is identical to condition ii) of (2.2.4) so the exact argument as in (2.5.3) can be repeated. ∎

## 3. Notes and Remarks

Just as these notes were being written, Wagner [64] obtained similar results using related but different methods. In the case of subspaces, the results in [64] are identical to the results in section 2 of Chapter III. However for quotient spaces Wagner obtains more. In particular, the requirement that $\alpha$ be multiplicately stable and sublinear in (2.2.4) is replaced by the weaker requirement that $\alpha$ be stable. More seriously, the condition on the diameter in (2.2.8) can be replaced by the same condition as in (2.2.7) so these results can be merged into a single characterization. Also, Vogt and Wagner [62], [63] generalize these results (in the case of infinite type power series spaces only) to quotient spaces without the assumption of a basis.

The main reason for including the present approach is to describe the duality between constructions of subspaces and constructions of quotient spaces. Thus we have tried to work as much as possible along the same lines as in Chapter III. This approach is useful in other problems. We will see a major example of it in VI (2.2). It seems that there is indeed some sort of duality between subspaces and quotient spaces, but sometimes, one has to work rather hard to discover it.

The material in sections 1 (2.1) and (2.2) is an attempt to generalize the work of the author and Robinson [34].

The material in section (2.3) is interesting because of a question which is suggested. If X is isomorphic to a subspace of Y and also to a quotient space of Y, is X in some sense of the same type as Y? Propositions (2.3.2) and (2.3.3) say that this is the case if Y is a power series space. But if Y is a countable product of (s) = $\Lambda_\infty(\log n)$ then by (2.4.2), $\Lambda_1(n)$, say, is a quotient space of (s) and by the theory of Kōmura and Kōmura [41], $\Lambda_1(n)$ is a subspace of Y. Thus we do not have this conclusion. It might be interesting to determine exactly when we do. This is closely related to the considerations of Chapter V.

The details of sections (2.4) and (2.5) are worked out here for the first time except that (2.4.3), (2.5.2) and (2.5.4) follow easily from the results of Zahariuta [65] and (2.4.1), (2.4.4) can be obtained easily from elementary considerations.

# CHAPTER V

# COMPLEMENTED SUBSPACES

## 1. Complemented Basic Sequences

(1.1) A basic sequence in a nuclear Fréchet space E is a complemented basic sequence if the subspace it generates is complemented in E. If E has a basis $(x_i)$ and $(y_i)$ is a subsequence of this basis, then because the basis is absolute [35] it follows immediately that $(y_i)$ is a complemented basic sequence. In this section we consider the possibility that this is the only way to construct a complemented basic sequence.

(1.2) Lemma

If $(x_n)$ is a basis in a nuclear Fréchet space E then $(x_n)$ can be reordered in such a way that for any seminorm $|\cdot|$ on E there is a seminorm $||\cdot||$ on E $\ni$

$$n^2|x_n| \leq ||x_n|| \quad \forall n.$$

Proof

By the nuclearity we may take an increasing sequence of seminorms $(||\cdot||_k)$ defining the topology on E $\ni$ if

$$c_n^k = \frac{||x_n||_k}{||x_n||_{k+1}} \quad (c_n^k = 0 \text{ if } ||x_n||_{k+1} = 0)$$

then $\sum_{n=1}^{\infty} c_n^k < \infty \ \forall k$. Now let $(k_n)$ be a strictly increasing sequence of indices such that for each n,

$$\sum_{m>k_n} c_m^j \leq \frac{1}{n^3} \text{ for } j = 1,\dots,n.$$

Then if we set $c_m = \max_{j<n} c_m^j$ for $k_n < m \leq k_{n+1}$ and $c_m = c_m^1$ for $m \leq k_n$ it follows that $c_m \geq c_m^j$ for $m > k_j$ so

$$\sum_{m=1}^{\infty} c_m = \sum_{m=1}^{k_1} c_m + \sum_{n=1}^{\infty} \sum_{m=k_n+1}^{k_{n+1}} c_m \leq \sum_{m=1}^{k_1} c_m + \sum_{n=1}^{\infty} \sum_{m>k_n} \max_{j\leq n} c_m^j$$

$$\leq \sum_{n=1}^{\infty} \sum_{m>k_n} \sum_{j=1}^{n} c_m^j = \sum_{n=1}^{\infty} \sum_{j=1}^{n} \sum_{m>k_n} c_m^j \leq \sum_{n=1}^{\infty} \frac{1}{n^2} < \infty .$$

Now we reorder $(x_n)$ so that $(c_n)$ is monotone and it follows that $B = \sup_n nc_n < \infty$ and, given $k$, $c_n^k \leq c_n$ for $n$ sufficiently large. Hence there exists $B_k \ni \sup_n nc_n^k = B_k < \infty$, which means that

$$n||x_n||_k \leq B_k||x_n||_{k+1} \quad \forall n,k.$$

Finally, given $|\cdot|$, choose $k \ni$ there is $C > 0 \ni |x| \leq C||x||_k$ $\forall x \in E$ and let $||\cdot|| = CB_kB_{k+1}||\cdot||_{k+2}$. Then, for each $n$,

$$|x_n| \leq C||x_n||_k \leq \frac{CB_kB_{k+1}}{n^2}||x_n||_{k+2} = \frac{1}{n^2}||x_n||. \quad ∎$$

(1.3) Theorem

Let $E$ be a nuclear Fréchet space with basis $(x_n)$ and let $(y_n)$ be a complemented basic sequence in $E$. Then for any representation $a$ of $(x_n)$ there exists a sequence $(t_n)$ of positive numbers and a sequence $(j_n)$ of indices with $\lim_n j_n = \infty \ni (y_n)$ has the representation $b$ where $b_n^k = t_n a_{j_n}^k$ $\forall n,k$.

<u>Proof</u>

First we assume that $(x_n)$ has been reordered $\ni$ the conclusion of (1.2) holds. This leads only to a reordering of $(j_n)$ and changes nothing.

Now let F be the space generated by $(y_n)$. Since F is complemented and $(y_n)$ is a basis, we have a sequence $(g_n)$ in $E' \ni g_n(y_m) = \delta_{nm}$ and for each $x \in E$, $(g_n(x) y_n)_n$ is a bounded sequence.

We can write

$$y_n = \sum_i \xi_i^n x_i, \quad \eta_i^n = g_n(x_i)$$

so that

$$1 = |g_n(y_n)| = |\sum_i \xi_i^n \eta_i^n| \leq \sum_i |\xi_i^n \eta_i^n| .$$

Let $j_n = \max\{j: \sup_i |\xi_i^n \eta_i^n| = |\xi_j^n \eta_j^n|\}$. Then if $\sigma = \sum_{n=1}^{\infty} \frac{1}{n^2}$ it follows that

$$\frac{1}{\sigma (j_n)^2} \leq |\xi_{j_n}^n \eta_{j_n}^n| \quad \forall n .$$

Set $t_n = \frac{1}{\eta_{j_n}^n}$. In order to establish the desired conclusions, let $(||\cdot||_k)$ be the sequence of sup norms in E relative to the matrix a. By (1.2), given k we have $\ell \ni$

$$\frac{1}{\sigma} t_n a_{j_n}^k \leq (j_n)^2 |\xi_{j_n}^n| a_{j_n}^k \leq |\xi_{j_n}^n| a_{j_n}^{\ell} \leq ||y_n||_{\ell} .$$

On the other hand, by the uniform boundedness principle, given k we have $\ell$ and $C > 0 \ni |g_n(x)| \, ||y_n||_k \leq C ||x||_{\ell}$ $\forall n,x$ and so

$$||y_n||_k \leq C t_n a_{j_n}^{\ell} .$$

Thus b is a representation of $(y_n)$ and it then follows from the nuclearity that $\lim_n j_n = \infty$. ∎

(1.4) The previous result comes close to showing that the simple construction in (1.1) actually gives all complemented basic sequences. It would only be necessary to refine the proof of (1.3) so as to be able to choose $(j_n)$ so that all terms are distinct. We are led to the following problem.

Conjecture. If X,Y are nuclear Fréchet spaces with bases, then Y is isomorphic to a complemented subspace of X iff Y is isomorphic to the space generated by a subsequence of the basis in X.

(1.5) There are some simple cases in which the conjecture of (1.4) can be established.

(1.5.1) Proposition

If, in the context of (1.4), Y is isomorphic to the space generated by a block basic sequence wrt the basis in X, then the conjecture holds.

Proof

In the proof of (1.3) it is clear that $j_n$ must lie in the block determined by $y_n$ (otherwise $\xi^n_{j_n} = 0$) and so $j_n \neq j_m$ when $n \neq m$. ∎

(1.5.2) Proposition

The conjecture holds for any stable power series space.

Proof

Let $\alpha$ be the stable exponent sequence and let $(j_n)$ be the sequence from (1.3). We may reorder the basic sequence $(y_n)$ so that $j_n \leq j_{n+1}$ and then if we set $\beta_n = \alpha_{j_n+n}$ it follows that $j_n + n < j_{n+1} + n+1$. On the other hand, $\beta$ defines the same space as $(\alpha_{j_n})$ because, in view of III (2.4.2) and the stability, we have constants C, A $\ni$

$$\alpha_{j_n} \leq \alpha_{j_n+n} \leq \alpha_{2j_n} + \alpha_{2n} \leq A(\alpha_{j_n}+\alpha_n) \leq A(\alpha_{j_n}+C\alpha_{j_n}) = A(1+C)\alpha_{j_n},$$

so the result follows. ∎

## 2. Bases in Complemented Subspaces

(2.1) The deepest results concerning the conjecture of (1.4) have been obtained in connection with the following:

Problem. Does every complemented subspace of a nuclear Fréchet space with basis have a basis?

The methods involved in this investigation are quite different from the methods which have been described so far in these notes. They present a good transition to the considerations of the next chapter. We present one general result, work it out for certain Köthe spaces and give some applications.

(2.2) General Result

(2.2.1) We assume throughout that E, F, X are nuclear Fréchet spaces whose topologies are given by the increasing sequences of norms, $(||\cdot||_k)$, $(|\cdot|_k)$, $(|||\cdot|||_k)$ respectively. We denote by $E_k$, $F_k$, $X_k$ the completion of the normed spaces $(E,||\cdot||_k)$, $(F,|\cdot|_k)$, $(X,|||\cdot|||_k)$ If $k \leq j$ then $E_j \to E_k$ will denote the canonical map and similarly for $F_k$, $X_k$, E, F, X.

We assume that $(x_n)$ is a basis for E, a is a representation of $(x_n)$ and

$$||x||_k = (\sum_n (|\xi_n|a_n^k)^2)^{\frac{1}{2}}, \quad x = \sum_n \xi_n x_n \in E.$$

We assume that X is a subspace of E with injection $q: X \to E$ and $|||x|||_k = ||qx||_k \ \forall k, \ x \in X$.

We assume that $\pi: F \to X$ is a quotient map.

In general, if $|\cdot|$ is a norm on a vector space L, we will denote by $|\cdot|^*$ the dual norm on $(L,|\cdot|)^*$. If u, v, w are norms on L and $t \geq 0$ we shall define

$$M(u,v,w;t) = \sup\{v(x): x\in L, \ w(x) \leq 1, \ u(x) \leq t\} .$$

(2.2.2) We will consider the following two conditions which may or may not hold.

i) There is a dense subspace $F_\infty$ of $F$ and a norm $|\cdot|_\infty$ on $F_\infty$ making it a separable Hilbert space and

$$\sup_k |y|_k \leq |y|_\infty, \quad y \in F_\infty$$

$$\inf_k |\psi|_k^* \leq |\psi|_\infty^*, \quad \psi \in \bigcup_k F_k^* \quad .$$

ii) There exists $k_1 \in \mathbb{N} \ni$ if $\ell_1 \in \mathbb{N}$, $k \geq k_1$ and $(a_n)$ is any decreasing sequence of positive numbers $\ni \forall j \geq k_1$ there exists $m \in \mathbb{N} \ni$

$$a_{n+m} \leq d_n(E_j \to E_{k_1}) \quad \forall n$$

then there exists $\ell \geq \ell_1$ and $\bar{k} \geq k \ni \forall A,B > 0$ we have

$$\sum_n M(||\cdot||_{k_1}, ||\cdot||_k, ||\cdot||_{\bar{k}}; \frac{a_n}{B}) \sup_{\bar{\ell}} M(|\cdot|_{\ell_1}^*, |\cdot|_\ell^*, |\cdot|_{\bar{\ell}}^*; \frac{A}{a_n}) < \infty \ .$$

(2.2.3) We fix a sequence $(y_i)$ in $F_\infty \ni$ its linear span $F_\infty^o$ is dense in $F_\infty$ and we extend the set $\{q\pi(y_i), x_i: i = 1,2,...\}$ to a sequence $(u_n)$ in $E \ni$ its linear span $E_\infty^o$ is dense in $E$. Finally we set $X_o = \pi(F_\infty^o)$ so that $X_o$ is dense in $X$ and $X_o \subset q^{-1}(E_\infty^o)$.

(2.2.4) It is clear from the definitions that we have increasing functions $\nu, \mu: \mathbb{N} \to \mathbb{N}$ and sequences $(A_k)$, $(B_k)$ of positive numbers $\ni$

$$||q\pi(y)||_k \leq A_k |y|_{\nu(k)} \quad \forall k \in \mathbb{N}, \ y \in F$$

$$x \in X, |||x|||_{\mu(k)} \leq 1 \Rightarrow \exists\, y \in F \ni |y|_k \leq B_k \text{ and } \pi(y) = x \ .$$

(2.2.5) For each $k \in \mathbb{N}$ we choose $C_k \geq \max\{2^k A_k, \max_{n<k} 2^k ||u_n||_k\} \ni$

$$\sup_k \frac{2^k ||x||_k}{C_k} < \infty \quad \forall x \in E^o_\infty$$

so we may define $||\cdot||_\infty$ on $E^o_\infty$ by

$$||x||_\infty = (\sum_k (\frac{||x||_k}{C_k})^2)^{\frac{1}{2}} \quad , \ x \in E^o_\infty$$

and let $E_\infty$ be the completion of $(E^o_\infty, ||\cdot||_\infty)$. Hence,

$$||x||_k \leq C_k ||x||_\infty, \ x \in E^o_\infty, \quad k \in \mathbb{N}$$

so the inclusion $E^o_\infty \subset E$ extends to a map $J: E_\infty \to E$.

It follows immediately from (2.2.3), (2.2.4) that

$$||q\pi(y)||_\infty \leq |y|_\infty, \quad y \in F^o_\infty \quad .$$

(2.2.6) We define $|||\cdot|||_\infty$ on $X_o$ by $|||x|||_\infty = ||qx||_\infty$ and let $X_\infty$ be the completion of $(X_o, |||\cdot|||_\infty)$ and $\hat{q}: X_\infty \to E_\infty$ the extension of $q: X_o \to E^o_\infty$. It is an isometry.

If $k < \infty$, then by (2.2.4) and the definition of $|||.|||_k$ we have a unique map $\pi_k: F_{\nu(k)} \to X_k \ni \pi_k(x) = \pi(x)$, $x \in F$. It follows that $||\pi_k|| \leq A_k$.

If $k = \infty$ we use (2.2.5) to obtain $\pi_\infty: F_\infty \to X_\infty$ which agrees with $\pi$ and $||\pi_\infty|| \leq 1$.

(2.2.7) Proposition

If $E_\infty \to E_k$ is the composition of $J$ with $E \to E_k$ and

$$b_n = (\sum_k (\frac{a^k_n}{C_k})^2)^{\frac{1}{2}}$$

then $E_\infty \to E_k$ is 1-1 and $b_n < \infty$, $n,k \in \mathbb{N}$.

<u>Proof</u>

If $x \in E^o_\infty$ and $x = \sum_{n=1}^{\infty} \xi_n x_n$ we may write

$$\sum_n (|\xi_n| b_n)^2 = \sum_{n,k} \frac{1}{C_k^2}(|\xi_n| a_n^k)^2 = \sum_k \left(\frac{||x||_k}{C_k}\right)^2 = ||x||_\infty^2 < \infty .$$

Since $x_n \in E_\infty^o$ this implies that $b_n < \infty$.

This also permits us to define an isometry $U: E_\infty^o \to \ell_2$ by $U(x) = (b_n \xi_n)_n$, $x = \sum_n \xi_n x_n$. Since $x_n \in E_\infty^o$ it follows that $U(E_\infty^o)$ is dense in $\ell_2$ so U extends to an isometry onto, $U: E_\infty \to \ell_2$ and $U(x_n) = b_n e_n$ ($e_n$ = coordinate sequence in $\ell_2$).

Now we also have an isometry $V: E_k \to \ell_2 \ni V(x_n) = a_n^k e_n$. Define $W: \ell_2 \to \ell_2$ by $We_n = \frac{a_n^k}{b_n} e_n$ so that

$$Vo(E_\infty \to E_k) = WoU .$$

This is because, since U is an isomorphism, $(x_n)$ is total in $E_\infty$ and this equality holds on $x_n$. Finally, WoU is 1-1 so $E_\infty \to E_k$ is also. ∎

(2.2.8) Proposition

In the above context, if $\ell_1 \le \ell$ and $t \ge 0$ we have

i) $M(|||\cdot|||_{k_1}, |||\cdot|||_k, |||\cdot|||_\infty; t) \le M(||\cdot||_{k_1}, ||\cdot||_k, ||\cdot||_\infty; t)$

ii) $M(|\cdot|^*_{\ell_1}, |\cdot|^*_\ell, |\cdot|^*_\infty; t) \le \sup_\ell M(|\cdot|^*_{\ell_1}, |\cdot|^*_\ell, |\cdot|^*_\ell; t)$

iii) $M(|||\cdot|||^*_{k_1}, |||\cdot|||^*_{\mu(\ell)}, |||\cdot|||^*_\infty; t) \le$

$B_\ell M(|\cdot|^*_{\nu(k_1)}, |\cdot|^*_\ell, |\cdot|^*_\infty; A_{k_1} t)$

<u>Proof</u>

We obtain i) immediately from the fact that $|||x|||_\ell = ||qx||_\ell \ \forall \ell \le \infty$. To show ii), let $\varepsilon > 0$ and suppose $\psi \in F^*_{\ell_1} \subset F^*_\ell \subset F^*_\infty$. By (2.2.2) i) there is an $\bar{\ell} \ni |\psi|^*_{\bar{\ell}} \le (1+\varepsilon)|\psi|^*_\infty$. Hence if $|\psi|^*_\infty \le 1, |\psi|^*_{\ell_1} \le t$ then

$$\left|\frac{1}{1+\varepsilon}\psi\right|^*_{\bar{\ell}} \leq 1, \quad \left|\frac{1}{1+\varepsilon}\psi\right|^*_{\ell_1} \leq t$$

so

$$\left|\frac{1}{1+\varepsilon}\psi\right|^*_{\ell} \leq M(|\cdot|^*_{\ell_1}, |\cdot|^*_{\ell}, |\cdot|^*_{\ell}; t) \leq \sup_{\bar{\ell}} M(|\cdot|^*_{\ell_1}, |\cdot|^*_{\ell}, |\cdot|^*_{\ell}; t)$$

from which ii) follows.

Finally, suppose we have $\phi \ni |||\phi|||^*_{\infty} \leq 1, |||\phi|||^*_{k_1} \leq t$. Then using (2.2.6) we obtain,

$$|\pi^*_{k_1}(\phi)|^*_{\nu(k_1)} \leq ||\pi^*_{k_1}|| \ |||\phi|||^*_{k_1} \leq A_{k_1} t$$

$$|\pi^*_{k_1}(\phi)|^*_{\infty} = |\pi^*_{\infty}(\phi)|^*_{\infty} \leq |||\phi|||^*_{\infty} \leq 1.$$

Also, since X is dense in $X_{\mu(\ell)}$ it follows from (2.2.4) that

$$|||\phi|||^*_{\mu(\ell)} = |||\phi|_X|||^*_{\mu(\ell)} \leq B_{\ell}|\pi^*(\phi)|^*_{\ell} = B_{\ell}|\pi^*_{k_1}(\phi)|^*_{\ell}$$

and iii) follows from these three relations. ∎

(2.2.9) Theorem

In the above context, if the conditions of (2.2.2) hold then X has a basis.

Proof

Take $k_1$ from (2.2.2) ii). From (2.2.5), (2.2.6) we have for $k_1 \leq j$ and $x \in X_o$,

$$|||x|||_{k_1} \leq |||x|||_j \leq C_j|||x|||_{\infty}$$

so we have the commuting diagram,

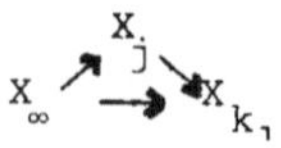

where the arrows represent the extensions of the identity. Thus by taking j large enough, since X is nuclear it follows that $X_\infty \to X_{k_1}$ is compact. Clearly it has dense range ($X_o$). It is 1-1 because otherwise we would have a sequence $(v_n)$ in $X_o$ which is $|||\cdot|||_\infty$-Cauchy and

$$|||v_n|||_\infty = 1, \ \lim_n |||v_n|||_{k_1} = 0 .$$

Now by (2.2.3) $qv_n \in E_\infty^o$ and $||qv_n||_\infty = 1, \ \lim_n ||qv_n||_{k_1} = 0$ and $(qv_n)$ is $||\cdot||_\infty$-Cauchy. By (2.2.7) this is impossible.

Thus we have a sequence $(e_n)$ which is an orthonormal basis for $X_\infty$ and an orthogonal basis for $X_{k_1}$. Moreover, if $a_n = |||e_n|||_{k_1}$ then we may assume that $(a_n)$ is decreasing. We will show that it satisfies the requirements of (2.2.2) ii).

Given $j \geq k_1$ choose $\ell \ni X_\ell \to X_j$ is compact. We have

$$|||x|||_\ell = ||q(x)||_\ell \leq C_\ell ||q(x)||_\infty = C_\ell |||x|||_\infty, \ x \in X_o$$

so $||X_\infty \to X_\ell|| \leq C_\ell$. Hence, choosing m so large that $C_\ell d_m(X_\ell \to X_j) \leq 1$ it follows from elementary properties of diameters that

$$a_n = d_n(X_\infty \to X_{k_1}) = d_n((X_\ell \to X_{k_1}) o (X_\infty \to X_\ell)) \leq C_\ell d_n(X_\ell \to X_{k_1})$$

$$= C_\ell d_n((X_j \to X_{k_1}) o (X_\ell \to X_j)) \leq d_{n-m}(X_j \to X_{k_1}) \quad .$$

Hence we may apply the conclusion of (2.2.2) ii). Let $\phi_n \in X_{k_1}^*$ be biorthogonal with $(e_n)$ in $X_{k_1}^*$. It follows that $|||\phi_n|||_{k_1}^* = \frac{1}{a_n}$, $|||\phi_n|||_\infty^* = 1$. We will complete the proof by showing that $\forall k$ there exists j and $p > 0 \ni$ if $x = \sum_n \phi_n(x) e_n$ is a finite sum, then

$$\sum_n |\phi_n(x)| \ |||e_n|||_k \leq p|||x|||_j \ .$$

Set $\ell_1 = \nu(k_1)$ and take any $k \geq k_1$. By (2.2.2) ii) we have $\ell \geq \ell_1$ and $\bar{k} \geq k$. Using (2.2.8), (2.2.5)

$$\sum_n |\phi_n(x)|\ |||e_n|||_k \leq \sum_n |||\phi_n|||^*_{\mu(\ell)} |||e_n|||_k |||x|||_{\mu(\ell)}$$

$$\leq \sum_n M(|||\cdot|||^*_{k_1}, |||\cdot|||^*_{\mu(\ell)}, |||\cdot|||^*_\infty; \frac{1}{a_n}) M(|||\cdot|||_{k_1}, |||\cdot|||_k, |||\cdot|||_\infty; a_n) |||x|||_{\mu(\ell)}$$

$$\leq B_\ell \sum_n M(|\cdot|^*_{\nu(k_1)}, |\cdot|^*_\ell, |\cdot|^*_\infty, \frac{A_{k_1}}{a_n}) M(||\cdot||_{k_1}, ||\cdot||_k, ||\cdot||_\infty; a_n) |||x|||_{\mu(\ell)}$$

$$\leq B_\ell \sum_n M(||\cdot||_{k_1}, ||\cdot||_k, ||\cdot||_\infty; a_n) \sup_{\bar{\ell}} M(|\cdot|^*_{\ell_1}, |\cdot|^*_\ell, |\cdot|^*_{\bar{\ell}}; \frac{A_{k_1}}{a_n}) |||x|||_{\mu(\ell)}$$

$$\leq B_\ell \sum_n M(||\cdot||_{k_1}, ||\cdot||_k, \frac{1}{C_{\bar{k}}} ||\cdot||_{\bar{k}}; a_n) \sup_{\bar{\ell}} M(|\cdot|^*_{\ell_1}, |\cdot|^*_\ell, |\cdot|^*_{\bar{\ell}}; \frac{A_{k_1}}{a_n}) |||x|||_{\mu(\ell)}$$

$$\leq B_\ell C_{\bar{k}} \sum_n M(||\cdot||_{k_1}, ||\cdot||_k, ||\cdot||_{\bar{k}}; \frac{a_n}{C_{\bar{k}}}) \sup_{\bar{\ell}} M(|\cdot|^*_{\ell_1}, |\cdot|^*_\ell, |\cdot|^*_{\bar{\ell}}; \frac{A_{k_1}}{a_n}) |||x|||_{\mu(\ell)}$$

$$= p|||x|||_{\mu(\ell)}$$

and $p < \infty$ by (2.2.2) ii). ∎

## (2.3) Applications to Köthe spaces

### (2.3.1) Lemma

Let $\rho, \sigma, \lambda_i, \nu_i$, $i \in \mathbb{N}$ be positive numbers ∋

$$\lambda_{i+1} < \lambda_i, \nu_i < \nu_{i+1}, \ \lim_i \nu_i = \infty, \ \frac{\nu_i}{\lambda_i} < \frac{\nu_{i+1}}{\lambda_{i+1}} \text{ and } \frac{\nu_1}{\lambda_1} < \frac{\sigma}{\rho} .$$

Set

$$S = \max\{\sum_{i=1}^{\infty} u_i : u_i \geq 0, \sum_i \lambda_i u_i \leq \rho, \sum_i \nu_i u_i \leq \sigma\}$$

$$\gamma = \min\{i : \frac{\sigma}{\rho} < \frac{\nu_i}{\lambda_i}\}$$

Define p,q,r,s by the conditions $p + q = r + s = 1$, $0 < p$, $0 < r \leq 1$ and

$$\frac{\sigma}{\rho} = p \frac{\nu_{\gamma-1}}{\lambda_{\gamma-1}} + q\frac{\nu_\gamma}{\lambda_\gamma}, \ \frac{\rho}{\sigma} = r \frac{\lambda_{\gamma-1}}{\nu_{\gamma-1}} + s \frac{\lambda_\gamma}{\nu_\gamma} .$$

Then,

$$\max(\frac{\rho}{\lambda_{\gamma-1}}, \frac{\sigma}{\nu_\gamma}) \leq S \leq 2 \min\{\rho(\frac{p}{\lambda_{\gamma-1}} + \frac{q}{\lambda_\gamma}), \ \sigma(\frac{r}{\nu_{\gamma-1}} + \frac{s}{\nu_\gamma})\} .$$

Proof

We will make our computations under the assumption that for $i < j < \ell$,

$$\text{Det} \begin{pmatrix} 1 & 1 & 1 \\ \lambda_i & \lambda_j & \lambda_\ell \\ \nu_i & \nu_j & \nu_\ell \end{pmatrix} \neq 0 .$$

If this is not the case we can perturb the sequences $(\lambda_i)$, $(\nu_i)$ so that it does hold, establish our inequality and then let the perturbation go to 0 to establish the result.

Now let $W = \{u = (u_i): u_i \geq 0 \forall i, \sum_i \lambda_i u_i \leq \rho, \sum_i \nu_i u_i \leq \sigma\}$.

This is coordinatewise closed and closed in $\ell_1$ and

$$\limsup_{n\to\infty} \sup_{u\in W} \sum_{i=n}^{\infty} |u_i| = \limsup_{n\to\infty} \sup_{u\in W} \sum_{i=n}^{\infty} \frac{1}{\nu_i}\nu_i u_i \leq \sigma \lim_n \frac{1}{\nu_n} = 0,$$

so W is compact in $\ell_1$. The function $u \rightsquigarrow \sum_i u_i$ is continuous on $\ell_1$ so it attains its maximum on W at some $u\epsilon W$. That is, $S = \sum_i u_i$.

Suppose we have $i < j < \ell$ with $u_i, u_j, u_\ell > 0$. Then by our assumption on the determinant, we can find numbers $h_i, h_j, h_\ell \ni$

$$h_i + h_j + h_\ell > 0, \; \lambda_i h_i + \lambda_j h_j + \lambda_\ell h_\ell = \nu_i h_i + \nu_j h_j + \nu_\ell h_\ell = 0.$$

These relations continue to hold if $h_i, h_j, h_\ell$ are each multiplied by $\delta > 0$.

Thus if v is obtained from u by replacing $u_i, u_j, u_\ell$ by $u_i + \delta h_i$, $u_j + \delta h_j$, $u_\ell + \delta h_\ell$ and leaving the other coordinates unchanged, we obtain a larger value for S which is a contradiction.

Thus $S = \sum_i v_i$ where $v = (v_i) \in W$ and $v_i \neq 0$ for at most two values of i.

Let $W_k = \{u \in W: u_i \neq 0$ for exactly k values of $i\}$, $S_k = \max\{\sum_i u_i: u \in W_k\}$, and it follows that we need only check our inequality for $k = 1,2$. Clearly,

$$S_1 = \max_i \min\{\frac{\rho}{\lambda_i}, \frac{\sigma}{\nu_i}\} = \max\{\max_{i<\gamma} \min\{\frac{\rho}{\lambda_i}, \frac{\sigma}{\nu_i}\}, \max_{i\geq\gamma} \min\{\frac{\rho}{\lambda_i}, \frac{\sigma}{\nu_i}\}$$

$$= \max\{\max_{i<\gamma} \frac{\rho}{\lambda_i}, \max_{i\leq\gamma} \frac{\sigma}{\nu_i}\} = \max\{\frac{\rho}{\lambda_{\gamma-1}}, \frac{\sigma}{\nu_\gamma}\} .$$

This establishes the left inequality for $S_1$ and hence for S. To obtain the right inequality for $S_1$ we check,

$$\frac{\rho}{\lambda_{\gamma-1}} \leq \rho\left(\frac{p}{\lambda_{\gamma-1}} + \frac{q}{\lambda_\gamma}\right) \qquad \frac{\sigma}{\nu_\gamma} \leq \sigma\left(\frac{r}{\nu_{\gamma-1}} + \frac{s}{\nu_\gamma}\right)$$

$$\frac{\rho}{\lambda_{\gamma-1}} = \frac{\sigma}{\lambda_{\gamma-1}}\left(r\frac{\lambda_{\gamma-1}}{\nu_{\gamma-1}} + s\frac{\lambda_\gamma}{\nu_\gamma}\right) = \sigma\left(\frac{r}{\nu_{\gamma-1}} + \frac{s}{\nu_\gamma}\frac{\lambda_\gamma}{\lambda_{\gamma-1}}\right) \leq \sigma\left(\frac{r}{\nu_{\gamma-1}} + \frac{s}{\nu_\gamma}\right)$$

$$\frac{\sigma}{\nu_\gamma} = \frac{\rho}{\nu_\gamma}\left(p\frac{\nu_{\gamma-1}}{\lambda_{\gamma-1}} + q\frac{\nu_\gamma}{\lambda_\gamma}\right) = \rho\left(\frac{p}{\lambda_{\gamma-1}}\frac{\nu_{\gamma-1}}{\nu_\gamma} + \frac{q}{\lambda_\gamma}\right) \leq \rho\left(\frac{p}{\lambda_{\gamma-1}} + \frac{q}{\lambda_\gamma}\right) \quad .$$

Thus it remains only to establish the right inequality for $S_2$. Let $u \in W_2$ and suppose that $S = u_i + u_j$, $i < j$. Then

$$\lambda_i u_i + \lambda_j u_j \leq \rho \quad \text{and} \quad \nu_i u_i + \nu_j u_j \leq \sigma.$$

If both of these inequalities were strict we could increase S by slightly increasing $u_i$, $u_j$. Hence one is an equality. Suppose for example that

$$\lambda_i u_i + \lambda_j u_j < \rho \quad \text{and} \quad \nu_i u_i + \nu_j u_j = \sigma.$$

Since $\nu_i < \nu_j$, then $h_i + h_j > 0$ whenever $\nu_i h_i + \nu_j h_j = 0$ and $h_i > 0$. Hence for $h_i$, $h_j$ sufficiently small, we can replace $u_i$, $u_j$ by $u_i + h_i$, $u_j + h_j$ and get a larger value for S. A similar argument works in the other mixed case so we may assume that

$$\lambda_i u_i + \lambda_j u_j = \rho \quad \text{and} \quad \nu_i u_i + \nu_j u_j = \sigma$$

or since $\lambda_i \nu_j - \lambda_j \nu_i > 0$,

$$u_i = \frac{\rho\nu_j - \sigma\lambda_j}{\lambda_i\nu_j - \lambda_j\nu_i}, \quad u_j = \frac{\sigma\lambda_i - \rho\nu_i}{\lambda_i\nu_j - \lambda_j\nu_i} .$$

But $u_i$, $u_j > 0$ so we have

$$\frac{\nu_i}{\lambda_i} < \frac{\sigma}{\rho} < \frac{\nu_j}{\lambda_j}$$

so $i < \gamma \le j$. Hence we may compute

$$u_i + u_j = \frac{\rho(\nu_j - \nu_i) + \sigma(\lambda_i - \lambda_j)}{\lambda_i\nu_j - \lambda_j\nu_i} = \sigma\,\frac{\frac{\rho}{\sigma}(\nu_j - \nu_i) + (\lambda_i - \lambda_j)}{\lambda_i\nu_j - \lambda_j\nu_i}$$

$$= \sigma\,\frac{r\frac{\lambda_{\gamma-1}}{\nu_{\gamma-1}}(\nu_j - \nu_i) + s\frac{\lambda_\gamma}{\nu_\gamma}(\nu_j - \nu_i) + r(\lambda_i - \lambda_j) + s(\lambda_i - \lambda_j)}{\lambda_i\nu_j - \lambda_j\nu_i}$$

$$= \frac{r\sigma}{\nu_{\gamma-1}}\,\frac{\lambda_{\gamma-1}(\nu_j - \nu_i) + \nu_{\gamma-1}(\lambda_i - \lambda_j)}{\lambda_i\nu_j - \lambda_j\nu_i} + \frac{s\sigma}{\nu_\gamma}\,\frac{\lambda_\gamma(\nu_j - \nu_i) + \nu_\gamma(\lambda_i - \lambda_j)}{\lambda_i\nu_j - \lambda_j\nu_i}$$

$$\le \frac{r\sigma}{\nu_{\gamma-1}}\,\frac{\lambda_i\nu_j - \lambda_j\nu_i + \lambda_i\nu_j - \lambda_j\nu_i}{\lambda_i\nu_j - \lambda_j\nu_i} + \frac{s\sigma}{\nu_\gamma}\,\frac{\lambda_i\nu_j - \lambda_j\nu_i + \lambda_i\nu_j - \lambda_j\nu_i}{\lambda_i\nu_j - \lambda_j\nu_i}$$

$$= 2\sigma\left(\frac{r}{\nu_{\gamma-1}} + \frac{s}{\nu_\gamma}\right)$$

and, analogously,

$$u_i + u_j = \rho\,\frac{(\nu_j - \nu_i) + \frac{\sigma}{\rho}(\lambda_i - \lambda_j)}{\lambda_i\nu_j - \lambda_j\nu_i} = \rho\,\frac{p(\nu_j - \nu_i) + q(\nu_j - \nu_i) + p\frac{\nu_{\gamma-1}}{\lambda_{\gamma-1}}(\lambda_i - \lambda_j) + q\frac{\nu_\gamma}{\lambda_\gamma}(\lambda_i - \lambda_j)}{\lambda_i\nu_j - \lambda_j\nu_i}$$

$$= \frac{p\rho}{\lambda_{\gamma-1}}\,\frac{\nu_{\gamma-1}(\lambda_i - \lambda_j) + \lambda_{\gamma-1}(\nu_j - \nu_i)}{\lambda_i\nu_j - \lambda_j\nu_i} + \frac{q\rho}{\lambda_\gamma}\,\frac{\nu_\gamma(\lambda_i - \lambda_j) + \lambda_\gamma(\nu_j - \nu_i)}{\lambda_i\nu_j - \lambda_j\nu_i}$$

$$\leq \frac{p\rho}{\lambda_{\gamma-1}} \frac{\lambda_i\nu_j-\lambda_j\nu_i+\lambda_i\nu_j-\lambda_j\nu_i}{\lambda_i\nu_j-\lambda_j\nu_i} + \frac{q\rho}{\lambda_\gamma} \frac{\lambda_i\nu_j-\lambda_j\nu_i+\lambda_i\nu_j-\lambda_j\nu_i}{\lambda_i\nu_j-\lambda_j\nu_i}$$

$$= 2\rho\left(\frac{p}{\lambda_{\gamma-1}} + \frac{q}{\lambda_\gamma}\right) \quad .$$

∎

(2.3.2) Proposition

Let K(a), K(b) be strictly regular Schwartz Köthe spaces. Let $(||\cdot||_k)$, $(|\cdot|_k)$ be the $\ell_2$-norms relative to the matrices a,b respectively.

i) If $0 < t < \frac{a_1^{k_1}}{a_1^{\overline{k}}}$ and $\lim_i \frac{a_i^{\overline{k}}}{a_i^k} = \infty$, then if $k_1 < k < \overline{k}$,

$$M(||\cdot||_{k_1}, ||\cdot||_k, ||\cdot||_{\overline{k}}; t)^2 \leq 2\left\{ r\left(\frac{a_{\gamma-1}^k}{a_{\gamma-1}^{\overline{k}}}\right)^2 + s\left(\frac{a_\gamma^k}{a_\gamma^{\overline{k}}}\right)^2 \right\}$$

where $\gamma$,r,s are determined by the relations,

$$\gamma = \min\{i: \frac{a_i^{k_1}}{a_1^{\overline{k}}} < t\}, \quad t^2 = r\left(\frac{a_{\gamma-1}^{k_1}}{a_{\gamma-1}^{\overline{k}}}\right)^2 + s\left(\frac{a_\gamma^{k_1}}{a_\gamma^{\overline{k}}}\right)^2 \quad 0<r\leq 1, \ s = 1 - r.$$

ii) If $\frac{b_1^{\overline{\ell}}}{b_1^{\ell_1}} < t$ and $\lim_i \frac{b_i^{\ell}}{b_i^{\ell_1}} = \infty$ then if $\ell_1 < \ell < \overline{\ell}$,

$$M(|\cdot|_{\ell_1}^*, |\cdot|_\ell^*, |\cdot|_{\overline{\ell}}^*; t)^2 \leq 2\left\{ p\left(\frac{b_{\gamma-1}^{\overline{\ell}}}{b_{\gamma-1}^{\ell}}\right)^2 + q\left(\frac{b_\gamma^{\overline{\ell}}}{b_\gamma^{\ell}}\right)^2 \right\}$$

where $\gamma$,p,q are determined by the relations,

$$\gamma = \min\{i: t < \frac{b_i^{\overline{\ell}}}{b_i^{\ell_1}}\}, \quad t^2 = p\left(\frac{b_{\gamma-1}^{\overline{\ell}}}{b_{\gamma-1}^{\ell_1}}\right)^2 + q\left(\frac{b_\gamma^{\overline{\ell}}}{b_\gamma^{\ell_1}}\right) \quad 0<p\leq 1, \ q = 1 - p.$$

Proof

i) Set $u_i = (a_i^k x_i)^2$ for $x \in \frac{1}{a^k}\ell_2$ and define

$$\rho = t^2, \ \sigma = 1, \ \lambda_i = \left(\frac{a_i^{k_1}}{a_i^k}\right)^2, \ \nu_i = \left(\frac{a_i^{\bar{k}}}{a_i^k}\right)^2 .$$

Then the hypotheses of (2.3.1) hold and $\gamma, r, s$ mean the same here as there so

$$M(||\cdot||_{k_1}, ||\cdot||_k, ||\cdot||_{\bar{k}}; t)^2 = \sup\{\sum_i u_i : \sum_i \left(\frac{a_i^{\bar{k}}}{a_i^k}\right)^2 u_i \le 1, \ \sum_i \left(\frac{a_i^{k_1}}{a_i^k}\right)^2 u_i \le t^2\}$$

$$= \sup\{\sum_i u_i : \sum_i \nu_i u_i \le \sigma, \ \sum_i \lambda_i u_i \le \rho\} = S$$

and the conclusion follows from (2.3.1).

ii) Similarly, we set $u_i = \left(\frac{y_i}{b_i^\ell}\right)^2$ for $y \in b^\ell \ell_2$ and define

$$\rho = 1, \ \sigma = t^2, \ \lambda_i = \left(\frac{b_i^\ell}{b_i^{\bar{\ell}}}\right)^2, \ \nu_i = \left(\frac{b_i^\ell}{b_i^{\ell_1}}\right)^2 .$$

Then again the hypotheses of (2.3.1) hold and $\gamma, p, q$ mean the same here as there so

$$M(|\cdot|^*_{\ell_1}, |\cdot|^*_{\ell}, |\cdot|^*_{\bar{\ell}}, t)^2 = \sup\{\sum_i u_i = \sum_i \left(\frac{b_i^\ell}{b_i^{\bar{\ell}}}\right)^2 u_i \le 1, \ \sum_i \left(\frac{b_i^\ell}{b_i^{\ell_1}}\right)^2 u_i \le t^2\}$$

$$= \sup\{\sum_i u_i : \sum_i \lambda_i u_i \le \rho, \ \sum_i \nu_i u_i \le \sigma\} = S$$

so again the conclusion follows from (2.3.1). ∎

(2.3.3) Proposition

Let $K(b)$ be a Schwartz space with $b$ strictly regular and $(|\cdot|_k)$ the $\ell_2$-norms relative to $b$. Let $\ell_1 < \ell$ and suppose that $\lim_i \frac{b_i^{\ell}}{b_i^{\ell_1}} = \infty$. Assume further that $\lim_k b_n^k = b_n^{\infty} < \infty$. Then (2.2.2) i) holds and if $t > \frac{b_1^{\infty}}{b_1^{\ell_1}}$,

$$\sup_{\overline{\ell}} M(|\cdot|_{\ell_1}^*, |\cdot|_{\overline{\ell}}^*, |\cdot|_{\ell}^*; t)^2 \leq 2\left\{p_\infty\left(\frac{b_{\gamma_\infty-1}^{\infty}}{b_{\gamma_\infty-1}^{\ell}}\right)^2 + q_\infty\left(\frac{b_{\gamma_\infty}^{\infty}}{b_{\gamma_\infty}^{\ell}}\right)^2\right\}$$

where $\gamma_\infty, p_\infty, q_\infty$ are determined by

$$\gamma_\infty = \min\{i: t < \frac{b_i^{\infty}}{b_i^{\ell_1}}\},\quad t^2 = p_\infty\left(\frac{b_{\gamma_\infty-1}^{\infty}}{b_{\ell_\infty-1}^{\ell_1}}\right)^2 + q_\infty\left(\frac{b_{\gamma_\infty}^{\infty}}{b_{\gamma_\infty}^{\ell_1}}\right)^2 \quad 0<p_\infty \leq 1,\ q_\infty = 1-p_\infty.$$

Proof

To establish (2.2.2) i) we take $F_\infty = \frac{1}{b^\infty}\ell_2$ and $|\cdot|_\infty$ to be defined by

$$|\xi|_\infty = (\sum_i |b_i^{\infty}\xi_i|^2)^{\frac{1}{2}} .$$

Recalling that $F_k = \frac{1}{b^k}\ell_2$ and $b_i^k \leq b_i^{k+1}$, the first relation in (2.2.2) i) is immediate. For the second, let $\eta \in F_{k_0}^* = b^{k_0}\ell_2$. Then, applying the dominated convergence theorem, we may compute,

$$(\inf_k |\eta|_k^*)^2 = \inf_k \sum_i \left|\frac{\eta_i}{b_i^k}\right|^2 = \lim_k \sum_i \left|\frac{\eta_i}{b_i^k}\right|^2 = \sum_i \left|\frac{\eta_i}{b_i^{\infty}}\right|^2 = (|\eta|_\infty^*)^2$$

as desired.

Turning to the second part of our proposition, we can define $\gamma_\ell$, $p_\ell$, $q_\ell$ for $\ell \in \mathbb{N}$ by

$$\gamma_\ell = \min\{i:\ t < \frac{b_i^\ell}{b_i^{\ell_1}}\},\quad t^2 = p_\ell\left(\frac{b^\ell_{\gamma_\ell - 1}}{b^{\ell_1}_{\gamma_\ell - 1}}\right)^2 + q_\ell\left(\frac{b^\ell_{\gamma_\ell}}{b^{\ell_1}_{\gamma_\ell}}\right)^2,\quad 0<p_\ell\le 1,\ q_\ell = 1-p_\ell$$

Then by (2.3.2) if $\ell < \bar{\ell}$,

$$M(|\cdot|^*_{\ell_1}, |\cdot|^*_\ell, |\cdot|^*_{\bar{\ell}}; t)^2 \le 2\left\{p_{\bar{\ell}}\left(\frac{b^{\bar{\ell}}_{\gamma_{\bar{\ell}}-1}}{b^{\ell}_{\gamma_{\bar{\ell}}-1}}\right)^2 + q_{\bar{\ell}}\left(\frac{b^{\bar{\ell}}_{\gamma_{\bar{\ell}}}}{b^{\ell}_{\gamma_{\bar{\ell}}}}\right)^2\right\} .$$

Clearly, $\gamma_{\bar{\ell}}$ decreases with $\bar{\ell}$ so it has a smallest value, $\tilde{\gamma}_\infty = \gamma_{\bar{\ell}}$ for $\bar{\ell}$ sufficiently large. Since M is decreasing in its third argument and $|\cdot|^*_{\bar{\ell}}$ decreases in $\bar{\ell}$, then M increases as $\bar{\ell}$ increases. Here $\sup_{\bar{\ell}}$ is actually $\underline{\lim}_{\bar{\ell}}$ and we may restrict consideration to $\bar{\ell}$ sufficiently large. Moreover, since $p_\ell$, $q_\ell$ are in [0,1] we may restrict consideration to a subsequence of $(\bar{\ell})$ on which $p_{\bar{\ell}}$, $q_{\bar{\ell}}$ converge to, say $\tilde{p}_\infty, \tilde{q}_\infty$ respectively. Hence, passing to the limit we obtain,

$$\sup_{\bar{\ell}} M(|\cdot|^*_{\ell_1}, |\cdot|^*_\ell, |\cdot|^*_{\bar{\ell}}; t) \le 2\left\{\tilde{p}_\infty\left(\frac{b^\infty_{\tilde{\gamma}_\infty - 1}}{b^\ell_{\tilde{\gamma}_\infty - 1}}\right)^2 + \tilde{q}_\infty\left(\frac{b^\infty_{\tilde{\gamma}_\infty}}{b^\ell_{\tilde{\gamma}_\infty}}\right)^2\right\} .$$

Now, for $\bar{\ell}$ sufficiently large,

$$\tilde{\gamma}_\infty = \gamma_{\bar{\ell}} = \min\{i: t < \frac{b_i^{\bar{\ell}}}{b_i^{\ell_1}}\} \ge \min\{i:\ t < \frac{b_i^\infty}{b_i^{\ell_1}}\} = \gamma_\infty .$$

On the other hand if $t < \frac{b_i^\infty}{b_i^{\ell_1}}$ then there exists $\bar{\ell} > \ell_1 \ni t < \frac{b_i^{\bar{\ell}}}{b_i^{\ell_1}}$ so that $\tilde{\gamma}_\infty \le \gamma_\infty$. Hence $\tilde{\gamma}_\infty = \gamma_\infty$.

Finally, if we pass to the limit in the relation determining $p_\ell$, $q_\ell$ along a sequence on which they converge to $\tilde{p}_\infty$, $\tilde{q}_\infty$, we obtain

$$\tilde{p}_\infty\left(\frac{b^\infty_{\gamma_\infty-1}}{b^{\ell_1}_{\gamma_\infty-1}}\right)^2 + \tilde{q}_\infty\left(\frac{b^\infty_{\gamma_\infty}}{b^{\ell_1}_{\gamma_\infty}}\right)^2 = t^2 = p_\infty\left(\frac{b^\infty_{\gamma_\infty-1}}{b^{\ell_1}_{\gamma_\infty-1}}\right)^2 + q_\infty\left(\frac{b^\infty_{\gamma_\infty}}{b^{\ell_1}_{\gamma_\infty}}\right)^2$$

and since distinct convex combinations of distinct numbers are distinct, it follows that $\tilde{p}_\infty = p_\infty$, $\tilde{q}_\infty = q_\infty$ so we have the desired inequality. ∎

(2.4) Main results

(2.4.1) Lemma

Let f be a Dragilev function which is three times differentiable and also satisfies

$$(f''(x))^2 \leq f'(x)f'''(x) + (f'(x))^2 f''(x), \quad x > 0$$

and let $0 \leq r, \delta \leq 1$, $x \geq 1$.

Then

$$re^{-f(\delta f^{-1}(\log x))} \leq e^{-f(\delta f^{-1}(\log\frac{x}{r}))} .$$

<u>Proof</u>

Let $F(x) = e^{f(x)}$. It follows by direct computation from our assumption that

$$F'''(x)F'(x) \geq (F''(x))^2, \quad x > 0$$

and hence $\frac{F''}{F'}$ is an increasing function so

$$F''(\tfrac{1}{\delta}x)F'(x) \geq F'(\tfrac{1}{\delta}x)F''(x), \quad x \geq 0$$

so the function

$$\frac{F'(\frac{1}{\delta}x)}{F'(x)}$$

is increasing and this remains so if x is replaced by $F^{-1}(x)$. Hence if $H(x) = F(\frac{1}{\delta}F^{-1}(x))$ it follows that $H'(x)$ is an increasing function and therefore,

$$\frac{H'(rH^{-1}(\frac{x}{r}))}{H'(H^{-1}(\frac{x}{r}))} \leq 1, \quad x \geq 1$$

and so we have

$$H(rH^{-1}(\frac{x}{r})) \leq x, \quad x \geq 1 .$$

But

$$H(x) = e^{f(\frac{1}{\delta}f^{-1}(\log x))}, \quad H^{-1}(x) = e^{f(\delta f^{-1}(\log x))}$$

which gives the desired result.

∎

(2.4.2) Theorem

Let f be a Dragilev function which satisfies the condition of (2.4.1) and let $E = L_f(\alpha,0)$, $F = L_f(\beta,0)$ and X isomorphic to a subspace of E and a quotient space of F. Then X has a basis.

Proof

We set $a_n^k = e^{-f(\frac{1}{k}\alpha_n)}$, $b_n^k = e^{-f(\frac{1}{k}\beta_n)}$ and $b_n^\infty = 1$ so the hypotheses of (2.3.2), (2.3.3) hold. Hence we have (2.2.2) i) so by (2.2.9) we have only to establish (2.2.2) ii).

We take R from I (5.4) and $k_1 = 1$. Now we have $\ell_1 \in \mathbb{N}$, $k \geq 1$ and a decreasing sequence of positive numbers $(a_n) \ni \forall j \geq 1$ there exists m $\ni$

$$a_{n+m} \leq e^{f(\frac{1}{j}\alpha_n)-f(\alpha_n)} \quad \forall n .$$

In view of (2.3.2) and (2.3.3) it is required to find $\ell \geq \ell_1$, $\bar{k} \geq k \ni$ if A, B > 0 then

$$\sum_n (X_n Y_n)^{\frac{1}{2}} < \infty$$

where

$$X_n = r\left(\frac{a^k_{\gamma-1}}{a^{\bar{k}}_{\gamma-1}}\right)^2 + s\left(\frac{a^k_{\gamma}}{a^{\bar{k}}_{\gamma}}\right)^2 \qquad Y_n = p\left(\frac{1}{b^{\ell}_{\nu-1}}\right)^2 + q\left(\frac{1}{b^{\ell}_{\nu}}\right)^2$$

and r, s, $\gamma$, p, q, $\nu$ are determined by

$$\gamma = \min\{i: \frac{a^1_i}{a^{\bar{k}}_i} < \frac{a_n}{B}\} \qquad \left(\frac{a_n}{B}\right)^2 = r\left(\frac{a^1_{\gamma-1}}{a^{\bar{k}}_{\gamma-1}}\right) + s\left(\frac{a^1_{\gamma}}{a^{\bar{k}}_{\gamma}}\right)^2$$

$$\nu = \min\{i: b_i^{\ell_1} < \frac{a_n}{A}\} \qquad \left(\frac{A}{a_n}\right)^2 = p\left(\frac{1}{b^{\ell_1}_{\nu-1}}\right) + q\left(\frac{1}{b^{\ell_1}_{\nu}}\right)^2$$

$$0 \leq p,\ q,\ r,\ s \leq 1, \quad p + q = r + s = 1.$$

For this purpose we take $\bar{k} > k$ and $\ell \ni \frac{4R\ell_1}{\ell} < \frac{1}{2kR}$. We write

$$\left(\frac{a_n}{B}\right)^2 = re^{2(f(\frac{1}{\bar{k}}\alpha_{\gamma-1})-f(\alpha_{\gamma-1})} + se^{2(f(\frac{1}{\bar{k}}\alpha_{\gamma})-f(\alpha_{\gamma}))} = r\tau_1^2 + s\tau_2^2,$$

so

$$\log\frac{1}{\tau_1} \leq f(\alpha_{\gamma-1}), \qquad \log\frac{1}{\tau_2} \leq f(\alpha_{\gamma})$$

so

$$\alpha_{\gamma-1} \geq f^{-1}(\log\frac{1}{\tau_1}), \qquad \alpha_{\gamma} \geq f^{-1}(\log\tau_2)\ .$$

So we may compute, for n sufficiently large by (2.4.1), and I (5.3),

(5.4),

$$X_n = re^{2(f(\frac{1}{\bar{k}}\alpha_{\gamma-1})-f(\frac{1}{k}\alpha_{\gamma-1}))} + se^{2(f(\frac{1}{\bar{k}}\alpha_{\gamma})-f(\frac{1}{k}\alpha_{\gamma}))}$$

$$\leq re^{-f(\frac{1}{k}\alpha_{\gamma-1})} + se^{-f(\frac{1}{k}\alpha_{\gamma})}$$

$$\leq re^{-f(\frac{1}{k}f^{-1}(\log\frac{1}{\tau_1}))} + se^{-f(\frac{1}{k}f^{-1}(\log\frac{1}{\tau_2}))}$$

$$\leq re^{-f(\frac{1}{2Rk}f^{-1}(\log\frac{1}{\tau_1^2}))} + se^{-f(\frac{1}{2Rk}f^{-1}(\log\frac{1}{\tau_2^2}))}$$

$$\leq e^{-f(\frac{1}{2Rk}f^{-1}(\log\frac{1}{r\tau_1^2}))} + e^{-f(\frac{1}{2Rk}f^{-1}(\log\frac{1}{s\tau_2^2}))}$$

$$\leq 2e^{-f(\frac{1}{2Rk}f^{-1}(\log(\frac{B}{a_n})^2)} .$$

On the other hand, we can compute analogously,

$$(\frac{A}{a_n})^2 = pe^{2f(\frac{1}{\ell_1}\beta_{\nu-1})} + qe^{2f(\frac{1}{\ell_1}\beta_{\nu})} = p\sigma_1^2 + q\sigma_2^2$$

so

$$\beta_{\nu-1} = \ell_1 f^{-1}(\log\sigma_1), \quad \beta_\nu = \ell_1 f^{-1}(\log\sigma_2)$$

so by (2.4.1),

$$Y_n = pe^{2f(\frac{1}{\ell}\beta_{\nu-1})} + qe^{2f(\frac{1}{\ell}\beta_{\nu})}$$

$$= pe^{2f(\frac{\ell_1}{\ell}f^{-1}(\log\sigma_1))} + qe^{2f(\frac{\ell_1}{\ell}f^{-1}(\log\sigma_2))}$$

$$\leq pe^{2f(\frac{\ell_1}{\ell}f^{-1}(\log\sigma_1^2))} + qe^{2f(\frac{\ell_1}{\ell}f^{-1}(\log\sigma_2^2))}$$

$$\leq e^{2f(\frac{\ell_1}{\ell}f^{-1}(\log p\sigma_1^2))} + e^{2f(\frac{\ell_1}{\ell}f^{-1}(\log q\sigma_2^2))}$$

$$\leq 2e^{2f(\frac{\ell_1}{\ell}f^{-1}(\log(\frac{A}{a_n})^2))} .$$

Using the fact that $\lim_n a_n = 0$ and I (5.3), (5.4) we have for n sufficiently large,

$$(X_nY_n)^{\frac{1}{2}} \leq 2e^{f(\frac{4R\ell_1}{\ell}f^{-1}(\log\frac{1}{a_n})-\frac{1}{2}f(\frac{1}{2kR}f^{-1}(\log\frac{1}{a_n}))}$$

$$\leq 2e^{-\frac{1}{4}f(\frac{1}{2kR}f^{-1}(\log\frac{1}{a_n}))} \leq 2e^{-\frac{1}{4}f(\frac{1}{2kR}\alpha_{n-m})}$$

which is summable since E is nuclear. ∎

(2.4.3) Theorem

Let $E = \Lambda_1(\alpha)$, $F = \Lambda_1(\beta)$ and X isomorphic to a subspace of E and a quotient space of F. Then X has a basis.

<u>Proof</u>

The first part of the proof of (2.4.2) may be repeated verbatim with f replaced by the identity except that now we must take $\ell > \ell_1 \ni \frac{1}{2k} - \frac{\ell_1}{\ell} = \delta > 0$ and $\bar{k} \geq 2k$. Then we reconsider the estimates for $X_n$, $Y_n$ as follows:

$$X_n = re^{2(\frac{1}{k}-\frac{1}{\bar{k}})\alpha_{\gamma-1}} + se^{2(\frac{1}{\bar{k}}-\frac{1}{k})\alpha_\gamma}$$

$$\leq re^{-\frac{1}{k}\alpha_{\gamma-1}} + se^{-\frac{1}{k}\alpha_\gamma} \leq re^{-\frac{1}{k}\log\frac{1}{\tau_1}} + se^{-\frac{1}{k}\log\frac{1}{\tau_2}}$$

$$= r(\tau_1)^{\frac{1}{k}} + s(\tau_2)^{\frac{1}{k}} \leq (r\tau_1^2)^{\frac{1}{2k}} + (s\tau_2^2)^{\frac{1}{2k}} \leq 2(\frac{a_n}{B})^{\frac{1}{k}}$$

and

$$Y_n = pe^{\frac{2}{\ell}\beta_{\nu-1}} + qe^{\frac{2}{\ell}\beta_\nu} = pe^{\frac{2\ell_1}{\ell}\log\sigma_1} + qe^{\frac{2\ell_1}{\ell}\log\sigma_2}$$

$$= p(\sigma_1^2)^{\frac{\ell_1}{\ell}} + q(\sigma_2^2)^{\frac{\ell_1}{\ell}} \le (p\sigma_1^2)^{\frac{\ell_1}{\ell}} + (q\sigma_2^2)^{\frac{\ell_1}{\ell}} \le 2(\frac{A}{a_n})^{\frac{2\ell_1}{\ell}} .$$

Also, our assumption on $(a_n)$ implies that we have $m \ni$

$$a_n \le e^{-\frac{1}{2}\alpha_{n-m}} \quad \forall n$$

so we have $C > 0 \ni$

$$(X_n Y_n)^{\frac{1}{2}} \le C(a_n)^{\frac{1}{2k}-\frac{\ell_1}{\ell}} \le Ce^{-\frac{\delta}{2}\alpha_{n-m}}$$

and this is summable because $(\alpha_n)$ is a nuclear exponent sequence of finite type. ∎

(2.5) Some consequences

(2.5.1) Theorem

If E is a finite type power series space or $L_f(\alpha,0)$ where f is a Dragilev function satisfying the hypothesis of (2.4.1), then every complemented subspace of E has a basis.

Proof

Immediate from (2.4.2), (2.4.3). ∎

(2.5.2) Theorem

If E is a finite type power series space or $L_f(\alpha,0)$ where f is a Dragilev function satisfying the hypothesis of (2.4.1), then the conjecture of (1.4) holds for complemented subspaces of E.

Proof

Let X be a complemented subspace of E and let Y be a complement. By (2.4.2), (2.4.3) X and Y have bases, say $(x_n)$, $(y_n)$. Since E is isomorphic to X×Y the union of $(x_n)$, $(y_n)$ is a basis for E, say $(z_j)$. Now given any basis for E, $(u_j)$ it follows that $(z_j)$ is quasi-equivalent to $(u_j)$, since E has the quasi-equivalence property. Hence $(x_n)$, being a subsequence of $(z_j)$ is quasi-equivalent to a subsequence of $(u_j)$. That is, X is isomorphic to the space generated by a subsequence of $(u_j)$. ∎

3. Notes and Remarks

The material in section 1 is due to Bessaga [4].

The material in section 2 is an attempt to generalize Theorem (2.4.3) which is due to Mitiagin [46]. (See also Mitiagin and Henkin [49]). In that work the proof is based on interpolation theory. Our proof of Theorem (2.2.9) follows the same lines except that the interpolation inequality is replaced by the assumption in (2.2.2) ii) which then must be verified directly.

The generalization is proper because of Theorem (2.4.2) and the easily checked fact that $f(x) = e^x-1$ is a Dragilev function which satisfies the condition of (2.4.1). It seems, however, that this method will not be helpful in case $E = L_f(\alpha,r)$, $r \neq 0$.

In [49] it is actually shown that if X is isomorphic to a subspace of $\Lambda_1(\alpha)$ and a quotient space of $\Lambda_1(\alpha)$ then X is isomorphic to a complemented subspace of $\Lambda_1(\alpha)$. Such a conclusion might also follow in our context from a more careful analysis of the estimates which appear in our arguments.

It might be interesting to observe that in the proof of (2.2.7) the fact that E has a basis is used in an essential way to show that $E_\infty \to E_k$ is 1-1. This has ramifications in other contexts.

In Theorem (2.5.2), the case E is a finite type power series space is due to Mitiagin [46] and case $E = L_f(\alpha,0)$ appears here for the first time.

Using methods entirely unrelated to the ideas in this work Mitiagin [47] has shown that the conjecture of (1.4) holds when E is an infinite type power series space.

## CHAPTER VI

## APPROXIMATION PROPERTIES

### 1. Definitions and Problems

(1.1) The existence of a basis in a nuclear Fréchet space can be thought of as an approximation property-in the sense that an element of the space can be approximated by the partial sums of its expansion in terms of the basis. We may also consider other (usually weaker) approximation properties for a nuclear Fréchet space E.

(1.1.1) E has the approximation property (AP) if there is a net $(A_d)_{d\varepsilon D}$ of continuous linear operators $A_d$: $E \to E \ni \dim A_d(E) < \infty$ and $\lim_{d\varepsilon D} A_d x = x$, $x \varepsilon E$.

(1.1.2) E has the bounded approximation property (BAP) if the net in (1.1.1) can be chosen to be a sequence $(A_n)_{n\varepsilon \mathbb{N}}$.

(1.1.3) E has a finite dimensional decomposition (FDD) if there is a sequence $(A_n)_{n\varepsilon \mathbb{N}}$ of continuous linear operators $A_n$: $E \to E \ni$ $\dim A_n(E) < \infty$, $A_n A_m = \delta_{nm} A_n$ and $x = \sum_{n=1}^{\infty} A_n x$, $x \varepsilon E$.

(1.1.4) E has an r-finite dimensional decomposition (r-FDD), $r \varepsilon \mathbb{N}$ if the sequence $(A_n)$ in (1.1.3) can be chosen such that $\dim A_n(E) \leq r$.

(1.1.5) E has a strong finite dimensional decomposition (SFDD) if there is an $r \varepsilon \mathbb{N} \ni$ E has an r-FDD.

(1.1.6) E has an unconditional partition of the identity (UPI) if there is a sequence $(A_n)$ of continuous linear operators $A_n$: $E \to E \ni$ $\dim A_n(E) < \infty$ and $x = \sum_{n=1}^{\infty} A_n x$ with the convergence being unconditional, $x \varepsilon E$.

(1.1.7) E has an r-unconditional partition of the identity (r-UPI), $r \varepsilon \mathbb{N}$ if the sequence $(A_n)$ in (1.1.6) can be chosen such that $\dim A_n(E) \leq r$.

(1.1.8) E has a strong unconditional partition of unity (SUPI) if there is an $r \varepsilon \mathbb{N} \ni$ E has an r-UPI.

(1.2) We can state without proof some simple and/or standard properties of the above concepts.

(1.2.1) Every nuclear Fréchet space has AP.

(1.2.2) A 1-FDD is exactly the same as a basis.

(1.2.3) A complemented subspace of a space with a UPI (r-UPI) again has a UPI (r-UPI).

(1.2.4) An FDD on a nuclear Fréchet space is also a UPI.

(1.2.5) Let $(A_n)$ be a UPI in a nuclear Fréchet space E. Since E is nuclear it follows that $\sum_n ||A_n x|| < \infty$ for all $x \in E$ and each continuous seminorm $||\cdot||$ on E. Since E is barreled, the set $U = \{x: \sum_n ||A_n x|| \leq 1\}$ is a nbd of 0 in E. Hence if $(||\cdot||_k)$ is a fundamental system of nbds of 0 for E it follows that $\forall k$ there exists j and $C > 0 \ni$

$$\sum_n ||A_n x||_k \leq C||x||_j, \quad x \in E$$

and so

$$\sum_n |\langle A_n x, u^*\rangle| \leq C||u^*||_k^* ||x||_j, \quad u^* \in E', \ x \in E .$$

Here E' is the dual of E and $||u^*||_k^* = \sup\{|\langle x,u^*\rangle| : ||x||_k \leq 1\}$ is the dual norm of $||\cdot||_k$.

(1.3) There are three types of problems relative to the approximation properties that we will consider here. Let P stand for one of the properties (1.1.i), i = 1,...,8.

(1.3.1) Existence. Does every nuclear Fréchet have property P?

(1.3.2) Permanence. If E has property P and F is a subspace of E or a quotient space of E, does it follow that F has property P?

(1.3.3) Comparison. Which of the 256 implications between these properties hold?

(1.3.4) We will obtain a complete solution to the existence problem, and will give a considerable amount of information about the comparison and permanence problems.

## 2. Negative Solutions to the Existence and Permanence Problems

(2.1) For all of the properties in (1.1) except AP and BAP, the permanence problem (and hence the existence problem) has a negative solution in a very strong sense. That is, every nuclear Fréchet space (except, of course, $\omega$) has both a subspace and a quotient space without these properties.

We begin with a general construction.

(2.1.1) Let L be a vector space. A biorthogonal pair on L is a pair $(x_j,x_j^*)$, $j = 1,2,\dots$, where the index set may be finite or infinite, $x_j \in L$, $x_j^* \in L^*$ (the algebraic dual of L) and $\langle x_j,x_i^*\rangle = \delta_{ij}$. If $(t_{ij})$ is an invertible matrix of scalars of order r+1 and $(x_j,x_j^*)$, $j = 1,\dots,r+1$ is a biorthogonal pair on L, we define the transform of $(x_j,x_j^*)$ by $(t_{ij})$ to be the biorthogonal pair $(y_i,y_i^*)$ on L determined by the relations

$$y_i = \sum_{j=1}^{r+1} t_{ij}x_j, \quad x_i^* = \sum_{j=1}^{r+1} t_{ji}y_j^*, \qquad i = 1,\dots,r+1 .$$

Let $N = \{(\alpha,\beta,\gamma) \in \mathbb{N}^3 : 1 < \alpha < \beta < \gamma\}$.

(2.1.2) Lemma

Let L be a vector space, $(\tau_{ij})$ an invertible matrix of order r+1 such that $\tau_{r+1,j} \neq 0$, $j = 1,\dots,r+1$.

Then there exists M>0 such that if $(f_j,f_j^*)$, $j = 1,\dots,r+1$ is a biorthogonal pair on L and A: L → L is linear with dim A(L) $\leq$ r, it follows that

$$|\langle Af_1,f_1^*\rangle| \leq M \sum_{i\neq j} |\langle Af_i,f_j^*\rangle| + \sum_{i=1}^{r} |\langle Av_{r+1},v_i^*\rangle| \qquad (1)$$

where $(v_i,v_i^*)$ is the transform of $(f_j,f_j^*)$ by $(\tau_{ij})$.

Proof

Fix $L, r$ and $(\tau_{ij})$.

Take one biorthogonal pair $(f_j, f_j^*)$ on $L$ and let $F$ be the subspace generated by $(f_j)$. Define

$$K = \{A:F \to F: A \text{ is linear, } \dim A(F) \leq r \text{ and } \sum_{i,j} |\langle Af_i, f_j^*\rangle| = 1\}$$

The set $K$ is clearly compact as a subset of $(r+1)^2$-dimensional Euclidean space and if $h: K \to \mathbb{R}$ is defined by the right hand side of (1) it follows that $h$ is continuous. Hence $h$ attains its minimum on $K$.

Now if $h(A) = 0$, then $\langle Af_i, f_j^*\rangle = 0$ for $i \neq j$ and $\langle Av_{r+1}, v_i^*\rangle = 0$ for $i = 1,\ldots,r$ so there exists scalar $\lambda \ni Av_{r+1} = \lambda v_{r+1}$ and since $v_{r+1} = \sum_{i=1}^{r+1} \tau_{r+1,i} f_i$ we have for $j = 1,\ldots,r+1$

$$\lambda\tau_{r+1,j} = \lambda\langle v_{r+1}, f_j^*\rangle = \langle Av_{r+1}, f_j^*\rangle = \sum_{i=1}^{r+1} \tau_{r+1,i}\langle Af_i, f_j^*\rangle = \tau_{r+1,j}\langle Af_j, f_j^*\rangle$$

so $(\langle Af_i, f_j^*\rangle)$ is a scalar matrix. But this matrix is singular because $\dim A(F) \leq r$ and so $\lambda = 0$ which implies $A = 0$. But this is impossible if $A \in K$.

Hence $h$ has a positive minimum so we have $M \ni$ (1) holds for this $(f_j, f_j^*)$ provided $A(F) \subset F$.

In general, if $A:L \to L$ let $P$ be the projection of $L$ onto $F$ along $\bigcap_{j=1}^{r+1} \ker f_j^*$. Then if $\dim A(L) \leq r$, we have (1) for $PA|_F$. But each $f_j^*, v_j^*$ vanishes on $(1-P)(L)$ so in (1) we may replace $PA|_F$ by $A$.

Finally, let $(g_j, g_j^*)$, $j = 1,\ldots,r+1$ be any other biorthogonal pair on $L$. There is an automorphism $U$ of $L \ni Uf_j = g_j$ and $f_j^* = U^*g_j^*$ for $j = 1,\ldots,r+1$. It then follows immediately that if $\langle w_i, w_i^*\rangle$ is the transform of $(g_j, g_j^*)$ by $(\tau_{ij})$ then $Uv_j = w_j$, $v_j^* = U^*w_j^*$, $j = 1,\ldots,r+1$.

The result then follows by applying (1) to $U^{-1}AU$. ∎

(2.1.3) Lemma

Let X be a nuclear Fréchet space with topology determined by a sequence of seminorms $(||\cdot||_k)$. Let $(\tau_{ij})$ be as in (2.1.2). Suppose that for each $(\alpha,\beta,\gamma) \in N$ and $\varepsilon > 0$ there exists a biorthogonal pair $(f_j, f_j^*)$, $j = 1,\ldots,r+1$ on X ∋ each $f_j \in X'$, the dual of X and

$$\sum_{1\leq i<j\leq r+1} (||f_i^*||_1^*||f_j||_\alpha + ||f_j^*||_{\alpha+1}^*||f_i||_\beta) +$$

$$\sum_{i=1}^{r} ||v_i^*||_{\beta+1}^*||v_{r+1}||_\gamma \leq \varepsilon,$$

where $(v_i, v_i^*)$ is the transform of $(f_j, f_j^*)$ by $(\tau_{ij})$.

Then X does not have an r-UPI.

Proof

Let $(A_n)$ be an r-UPI for X. If we apply (1.2.5) three times we can find $(\alpha,\beta,\gamma) \in N$ and $C > 0$ ∋

$$\sum_n |\langle A_n x, u^*\rangle| \leq C||u^*||_k||x||_j, \quad u^* \in X',\ x \in X$$

where $(k,j)$ takes on the values $(1,\alpha)$, $(\alpha+1,\beta)$, $(\beta+1,\gamma)$. Let M be the constant of (2.1.2) corresponding to X, $(\tau_{ij})$ and set $\varepsilon = \frac{1}{2MC}$.

Applying our hypothesis and (2.1.2) we may compute,

$$1 = |\langle f_1, f_1^*\rangle| = |\sum_n \langle A_n f_1, f_1^*\rangle| \leq \sum_n |\langle A_n f_1, f_1^*\rangle|$$

$$\leq M\sum_n (\sum_{i<j} (|\langle A_n f_j, f_i^*\rangle| + |\langle A_n f_i, f_j^*\rangle|) + \sum_{i=1}^{r} |\langle A_n v_{r+1}, v_i^*\rangle|)$$

$$\leq CM\Big(\sum_{i<j}(||f_i^*||_1^*||f_j||_\alpha + ||f_j^*||_{\alpha+1}^*||f_i||_\beta) + \sum_{i=1}^{r}||v_i^*||_{\beta+1}^*||v_{r+1}||_\gamma\Big)$$

$$\leq CM\varepsilon = \frac{1}{2}$$

which is a contradiction. ∎

(2.1.4) Lemma (selection lemma)

Let L be a vector space and $(\phi_n)$ a linearly independent sequence in L. Let $|\cdot|_k$, k = 1,...,8 be norms on the algebraic closure $L_o$ of $(\phi_n)$ ∍

a) $|x|_k = \sup_n |\xi_n||\phi_n|_k, \quad x = \sum_n \xi_n\phi_n \in L_o$

b) $\lim_n \dfrac{|\phi_n|_k}{|\phi_n|_{k+1}} = 0, \quad k = 1,\ldots,7$

Let $(t_{ij})$ be an invertible matrix of order r+1 and let $\delta > 0$.

Then there exist $u_j$, $x_j$, $y_j$, $z_j$, j = 1,...,r+1 distinct vectors taken from $(\phi_n)$, vectors $e_j$, $w_j$, j = 1,...,r+1 in the space generated by these 4r + 4 vectors and scalars $a_j$, $b_j$, $c_j$, $d_j$, j=1,..., r+1 ∍

i) $w_i = \sum_j t_{ij}e_j$, i = 1,...,r+1

ii) $e_j \rightsquigarrow a_ju_j$, j = 1,...,r+1 defines a $|\cdot|_k$-isometry for k = 1,2

iii) $e_j \rightsquigarrow b_jx_j$, j = 1,...,r+1 defines a $|\cdot|_k$-isometry for k = 3,4

iv) $w_j \rightsquigarrow c_jy_j$, j = 1,...,r+1 defines a $|\cdot|_k$-isometry for k = 5,6

v) $w_j \rightsquigarrow d_jz_j$, j = 1,...,r+1 defines a $|\cdot|_k$-isometry for k = 7,8 and for $1 \leq i < j \leq r+1$,

vi) $|e_j|_2 \leq \delta |e_i|_1$

vii) $|e_i|_4 \leq \delta |e_j|_3$

viii) $|w_j|_6 \leq \delta |w_i|_5$

ix) $|w_i|_8 \leq \delta |w_j|_7$

<u>Proof</u>

Let $(s_{ij})$ be the inverse of $(t_{ij})$. Once we have selected all $u_j$, $x_j$, $y_j$, $z_j$ and $a_j$, $b_j$, $c_j$, $d_j$ we will write for $j = 1,\ldots,r+1$,

$$e_j = a_j u_j + b_j x_j + \sum_i s_{ji}(c_i y_i + d_i z_i)$$

$$w_j = \sum_i t_{ji}(a_i u_i + b_i x_i) + c_j y_j + d_j z_j .$$

This will guarantee i).

Next we describe how the selection is made so as to obtain the isometries in ii),...,v). Let $\theta$ be a vector in the space generated by $e_1,\ldots,e_{r+1}$ so we may write,

$$\theta = \sum_j \xi_j e_j = \sum_j \eta_j e_j$$

$$= \sum_j \xi_j a_j u_j + \sum_j \xi_j b_j x_j + \sum_j (\sum_i \xi_i s_{ij}) c_j y_j + \sum_j (\sum_i \xi_i s_{ij}) d_j z_j$$

$$= \sum_j (\sum_i \eta_i t_{ij}) a_j u_j + \sum_j (\sum_i \eta_i t_{ij}) b_j x_j + \sum_j \eta_j c_j y_j + \sum_j \eta_j d_j z_j .$$

According to a), $|\theta|_k$ is a maximum of terms in either of the above two representations. Thus the isometries will be obtained if we can make our selection so that this maximum occurs at one of the $u_j$ terms when $k = 1,2$; one of the $x_j$ terms when $k = 3,4$; one of the $y_j$ terms when $k = 5,6$; and one of the $z_j$ terms when $k = 7,8$.

To see how this behavior can be arranged, we first specify that the order of selection will be that we choose all $(u_j)$ then $(a_j)$, then all $(x_j,b_j)$ then all $(y_j,c_j)$ and finally all $(z_j,d_j)$. There is no loss of generality if we assume that $\sup_j|\xi_j|\leq 1$.

Now we suppose that $(u_j)$, $(a_j)$ have been selected. When choosing $(x_j)$, $(b_j)$ we must have, in view of the first representation,

$$|b_j||x_j|_k \leq A, \qquad B \leq |b_j||x_j|_\ell \qquad k = 1,2 \text{ and } \ell = 3,4 \qquad (2)$$

for $j = 1,\ldots,r+1$ where A,B are constants determined by the values of $(u_j)$, $(a_j)$. When choosing $(y_j)$, $(c_j)$ we must have, in view of the first representation,

$$|c_j||y_j|_k \leq C \qquad k = 1,2,3,4 \qquad (3)$$

and, in view of the second representation,

$$D \leq |c_j||y_j|_\ell \qquad \ell = 5,6 \qquad (4)$$

for $j = 1,\ldots,r+1$ where C,D are constants determined by the values of $(u_j)$, $(a_j)$, $(x_j)$, $(b_j)$. Note that this can be done independently of $(\xi_i)$, $(\eta_i)$ because of the inequalities,

$$|\sum_i \xi_i s_{ij}| \leq \sum_i |s_{ij}|, \quad |\sum_i \eta_i t_{ij}| = |\xi_j| \leq 1.$$

In a similar manner we see that in choosing $(z_j)$, $(d_j)$ we must have, for $j = 1,\ldots,r+1$

$$|d_j||z_j|_k \leq E, \quad F \leq |d_j||z_j|_\ell \qquad k = 1,2,3,4,5,6 \text{ and } \ell = 7,8 \qquad (5)$$

where E,F are constants determined by the values of $(u_j)$, $(a_j)$, $(x_j)$,

$(b_j)$, $(y_j)$, $(z_j)$.

These inequalities will guarantee the isometries and we now consider how they must be adjusted to obtain vi),...,ix). We may assume that the isometries hold. There is still no restriction on $(u_j)$ so these can be any r+1 distinct terms from $(\phi_n)$. In view of the isometries we can easily choose $(a_j)$ such that vi) holds.

Now we specify that $x_j$, $b_j$ will be chosen in the order $x_1, b_1, \ldots, x_{r+1}, b_{r+1}$. To obtain vii) it is only necessary to make $|b_j||x_j|_3$ larger than a quantity which depends on $x_1, b_1, \ldots, x_{j-1}, b_{j-1}$. This can be obtained by using (2) but with B replaced by a perhaps larger quantity $B_j$. This will guarantee vii).

Because of the form of viii) it is necessary to choose $y_j$, $c_j$ in the order $y_{r+1}, c_{r+1}, \ldots, y_1, c_1$ and then we can use (4) with D replaced by $D_j$.

Finally we obtain ix) in a similar manner, choosing $z_j$, $d_j$ in the order $z_1, d_1, \ldots, z_{r+1}, d_{r+1}$ and using (5) with F replaced by $F_j$.

To summarize, our lemma is proved provided we can make our selection in the order described so that the following inequalities hold for each $j = 1,\ldots,r+1$

$$\frac{B_j}{|x_j|_\ell} \le |b_j| \le \frac{A}{|x_j|_k} \qquad k = 1,2 \text{ and } \ell = 3,4$$

$$\frac{D_j}{|y_j|_\ell} \le |c_j| \le \frac{C}{|y_j|_k} \qquad k = 1,2,3,4 \text{ and } \ell = 5,6$$

$$\frac{F_j}{|z_j|_\ell} \le |d_j| \le \frac{E}{|z_j|_k} \qquad k = 1,2,3,4,5,6 \text{ and } \ell = 7,8 .$$

The scalars $b_j$, $c_j$, $d_j$ can be chosen to satisfy these inequalities provided the following quantities are sufficiently small:

$$\sup_{\substack{k=1,2\\ \ell=3,4}} \frac{|x_j|_k}{|x_j|_\ell}, \qquad \sup_{\substack{1\le k\le 4\\ \ell=5,6}} \frac{|y_j|_k}{|y_j|_\ell}, \qquad \sup_{\substack{1\le k\le 6\\ \ell=7,8}} \frac{|z_j|_k}{|z_j|_\ell}.$$

In view of b) we may choose $x_j$, $y_j$, $z_j$ distinct vectors from $(\phi_n)$ so that these quantities are arbitrarily small. ∎

(2.1.5) Theorem

Any infinite dimensional nuclear Fréchet space E not isomorphic to $\omega$ has a subspace which has no SUPI.

Proof

It is well known [9] that E has a subspace with a continuous norm. If this subspace has no SFDD we are finished. If it has one, then by taking one vector from the range of each projection we obtain a basis. Thus by I (3.2) we may assume that $E = K(a)$. We will denote by $(||\cdot||_k)$ the "sup norms" defined in I (3.4).

Fix $r \in \mathbb{N}$ and let $\Lambda = \Lambda(r)$ be an infinite subset of the coordinate vectors in E. For each $\nu = (\alpha,\beta,\gamma,n) \in \mathcal{N} \times \mathbb{N}$ we will define an (r+1)-dimensional subspace $X_\nu$ of the subspace of E generated by $\Lambda$. To do this we divide $\Lambda$ into infinitely many pairwise disjoint infinite subsets $\Lambda_\nu$, $\nu \in \mathcal{N} \times \mathbb{N}$. For each $\nu$ we apply (2.1.4) to E; $\Lambda_\nu$; the norms $||\cdot||_1$, $||\cdot||_\alpha$, $||\cdot||_{\alpha+1}$, $||\cdot||_\beta$, $||\cdot||_{\beta+1}$, $||\cdot||_\gamma$, $||\cdot||_{\gamma+1}$, $||\cdot||_{\gamma+2}$; any invertible matrix $(t_{ij})$ of order r+1 $\ni$ $t_{r+1,j} \neq 0$, $j = 1,\dots,r+1$; and $\delta = \frac{1}{n}$.

From (2.1.4) we have $e_j$, $w_j$, $j = 1,\dots,r+1$ and we take $X_\nu$ to be the space which they generate. We define X to be the subspace of E generated by $\bigcup_\nu X_\nu$.

Now we apply (2.1.3) to X; $(||\cdot||_k)$; $(\tau_{ij}) = (t_{ij})$. Given $(\alpha,\beta,\gamma) \in \mathcal{N}$ and $\varepsilon > 0$ we take $\nu = (\alpha,\beta,\gamma,n)$ with $n \geq \frac{(r+1)^2}{}$ and invoke (2.1.4). We set $f_j = e_j$ and choose $f_i^* \varepsilon X' \ni \langle f_j, f_i^* \rangle = \delta_{ij}$ and $f_i^*$ vanishes on each $X_\mu$, $\mu \neq \nu$, $i,j=1,\dots,r+1$. Then if $(v_j, v_j^*)$ is the transform of $(f_j, f^*)$ by $(\tau_{ij})$ it follows from (2.1.4) i) that $w_j = v_j$, $j=1,\dots,r+1$.

Because of the form of the isometries in (2.1.4) ii), iii), iv) and the definition of $f_j^*$, $v_j^*$ we can easily compute for $j = 1,\dots,r+1$,

$$||f_j^*||_k^* = \frac{1}{||f_j||_k} \qquad k = 1,\alpha,\alpha+1,\beta$$

$$||v_j^*||_k^* = \frac{1}{||v_j||_k} \qquad k = \beta+1,\gamma \ .$$

These, along with (2.1.4) vi), vii) and viii) give us the required inequality in the statement of (2.1.3) so it follows that X does not have an r-UPI.

Finally we do this for each $r \in \mathbb{N}$, choosing $\Lambda(r)$ pairwise disjoint, and we obtain a space $X(r)$ which does not have an r-UPI. If $\hat{X}$ is the subspace of E generated by $\bigcup_r X(r)$ then each $X(r)$ is a complemented subspace of $\hat{X}$. By (1.2.3), $\hat{X}$ fails to have an r-UPI no matter what is r, that is, $\hat{X}$ has no SUPI. ∎

(2.1.6) Theorem

Any nuclear Fréchet space E not isomorphic to $\omega$ has a quotient space which has no SUPI.

Proof

By II (3.1.3) E has a quotient space with a basis and a continuous norm. Thus by I (3.2) we may assume that $E = K(a)$. We will denote by $||\cdot||_k$ the "$\ell_1$-norms" on E given by $||\xi||_k = \sum_n |\xi_n| a_n^k$, $\xi \varepsilon E$. The dual norms, $||\cdot||_k^*$ are given by $||\eta||_k^* = \sup_n \frac{|\eta_n|}{a_n^k}$ where $\eta$ is a finite sequence in $E' = K(a)^{\times} = \{\eta\colon$ there exists $k \ni \sup_n \frac{|\eta_n|}{a_n^k} < \infty\}$.

Fix $r \in \mathbb{N}$ and let $\Lambda = \Lambda(r)$ be an infinite subset of the coordinate vectors in $E'$. For each $\nu = (\alpha,\beta,\gamma,n) \in \mathcal{N} \times N$ we will define an $(r+1)$-dimensional subspace $Y_\nu$ of the subspace of $E'$ generated by $\Lambda$. To do this we divide $\Lambda$ into infinitely many pairwise disjoint infinite subsets $\Lambda_\nu$, $\nu \in \mathcal{N}\times N$. For each $\nu$ we apply (2.1.4) to $E'$; $\Lambda_\nu$; the norms $||\cdot||^*_{\gamma+2}, ||\cdot||^*_{\gamma+1}, ||\cdot||^*_{\gamma}, ||\cdot||^*_{\beta+1}, ||\cdot||^*_{\beta}, ||\cdot||^*_{\alpha+1}, ||\cdot||^*_{\alpha}, ||\cdot||^*_{1}$; any invertible matrix $(t_{ij})$ of order $r+1 \ni t_{i,r+1} \neq 0$, $i = 1,\ldots,r+1$; and $\delta = \frac{1}{n}$.

From (2.1.4) we have $e_j$, $w_j$, $j = 1,\ldots,r+1$ and we take $Y_\nu$ to be the space which they generate. We define $Y$ to be the subspace of $E'$ (with its strong topology) generated by $\cup_\nu Y_\nu$. Since $Y$ is a subspace of $E'$ and $E$ is reflexive, then $X = Y'$ is a quotient of $E$ and, for each $k$, the quotient norm $||\cdot||_k$ on $X$ induced by the norm $||\cdot||_k$ of $E$ is equal to $||\cdot||^{**}_k$, the dual norm of the restriction to $Y$ of the dual norm $||\cdot||^*_k$.

Now we apply (2.1.3) to $X$; $||\cdot||_k$ on $X$; $(\tau_{ij}) = (t_{ji})$. Given $(\alpha,\beta,\gamma) \in \mathcal{N}$ and $\varepsilon > 0$ we take $\nu = (\alpha,\beta,\gamma,n)$ with $n \geq \frac{(r+1)2}{\varepsilon}$ and invoke (2.1.4). We set $f_j^* = w_j$ and choose $f_i \in X \ni \langle f_i, f_j^*\rangle = \delta_{ij}$ and $\langle f_i, f^*\rangle = 0$ for $f^* \in Y_\mu$, $\mu \neq \nu$, $i,j=1,\ldots,r+1$. Then if $(v_j, v_j^*)$ is the transform of $(f_j, f_j^*)$ by $(\tau_{ij})$ it follows from (2.1.4) i) and a simple computation that $e_j = v_j^*$, $j = 1,\ldots,r+1$.

Because of the form of the isometries in (2.1.4) iii), iv), v) and the definition of $v_j$, $f_j$ we can easily compute for $j = 1,\ldots,r+1$

$$||v_j||_k = \frac{1}{||v_j^*||^*_k} \qquad k = \gamma, \beta+1$$

$$||f_j||_k = \frac{\phantom{1}}{||f_j^*||^*_k} \qquad k = \beta, \alpha+1, \alpha, 1 .$$

These, along with (2.1.4) vii), viii) and ix) give us the required inequality in the statement of (2.1.3) so it follows that X does not have an r-UPI.

Finally, we do this for each $r \in \mathbb{N}$, choosing $\Lambda(r)$ pairwise disjoint and we obtain a space Y(r) whose dual X(r) does not have an r-UPI. If $\hat{Y}$ is the subspace of E' generated by $\bigcup_r Y(r)$ then the dual, $\hat{X}$ of $\hat{Y}$ is a quotient of E $\ni$ each X(r) is a complemented subspace of $\hat{X}$. By (1.2.3), $\hat{X}$ fails to have an r-UPI no matter what is r, that is, $\hat{X}$ has no SUPI. ∎

(2.2) Consequences

(2.2.1) Corollary

Every nuclear Fréchet space not isomorphic to $\omega$ has both a subspace and a quotient space without a basis, a SUPI, an r-UPI, a UPI, an SFDD, an r-FDD or an FDD.

Proof

Immediate from (1.2.4), (2.1.5) and (2.1.6). ∎

(2.2.2) Corollary

There exist nuclear Fréchet spaces without a basis, a SUPI, an r-UPI, a UPI, an SFDD, an r-FDD or an FDD.

Proof

Immediate from (2.2.1). ∎

(2.2.3) Corollary

There is a nuclear Fréchet space which has an (r+1)-UPI ((r+1)-FDD) but has no r-UPI (r-FDD). Such a space has a SUPI (SFDD) but no r-UPI (r-FDD).

Proof

It is clear (with the help of (1.2.4)) that the space X(r) constructed in (2.1.5) is such a space. ∎

(2.2.4) Corollary

There is a nuclear Fréchet space which has a UPI (FDD) but has no SUPI (SFDD).

Proof

It is clear (with the help of 1.2.4) that the space $\hat{X}$ constructed in (2.1.5) is such a space.

∎

## 3. The Bounded Approximation Property

(3.1) Countably normed spaces

(3.1.1) Let E be a Fréchet space which admits a continuous norm so that its topology may be defined by an increasing sequence of norms $(||\cdot||_k)$. Then the identity map $(E,||\cdot||_{k+1}) \to (E,||\cdot||_k)$ is continuous so it has a unique extension to the completions, $(E,||\cdot||_{k+1})^{\wedge} \to (E,||\cdot||_k)^{\wedge}$. These latter maps are called the canonical maps with respect to $(||\cdot||_k)$.

(3.1.2) It is not hard to construct examples in which a canonical map fails to be 1-1. We say that E is countably normed if the norms $(||\cdot||_k)$ can be chosen so that each canonical map is 1-1.

(3.1.3) It is easy to see that the condition of (3.1.2) is exactly the same as the requirement that if $(x_n)$ is a sequence in E which is Cauchy with respect to both $||\cdot||_k$ and $||\cdot||_j$, then $\lim_n ||x_n||_k = 0$ iff $\lim_n ||x_n||_j = 0$.

(3.1.4) Using (3.1.3) it is obvious that a subspace of a countably normed space is again countably normed.

(3.1.5) Proposition

If E is a Fréchet space which admits a continuous norm and has the bounded approximation property, then there is a Fréchet space F which admits a continuous norm and has a FDD such that E is isomorphic to a subspace of F.

Proof

Let $(A_n)$ be the sequence of operators from the definition of BAP and $(||\cdot||_k)$ a sequence of norms defining the topology of E.

Let $B_1 = A_1$, $B_{n+1} = A_{n+1} - A_n$, $n \in \mathbb{N}$ so that each $B_n(E)$ is finite dimensional and $x = \sum_n B_n x$, $x \in E$.

We define a system of norms, $(|\cdot|_k)$ on E by

$$|x|_k = \sup_n ||\sum_{i=1}^{n} B_i x||_k, \qquad x \in E, \quad k \in \mathbb{N}.$$

Clearly $||x||_k \leq |x|_k$ and, on the other hand the unit balls of $|\cdot|_k$ are barrels in E and hence nbds of 0 in E. Thus $(|\cdot|_k)$ also defines the topology of E.

Now we define

$$F = \{(y_n): y_n \in B_n(E),\ n \in \mathbb{N} \text{ and } \sum_n y_n \text{ converges in E}\}$$

with topology given by the sequence of norms $(|\cdot|_k)$ where

$$|(y_n)|_k = \sup_n ||\sum_{i=1}^{n} y_i||_k \quad .$$

It is easy to see that F is a Fréchet space and the coordinate projections form an FDD. Finally, the map $x \rightsquigarrow (B_n x)$ is clearly an isomorphism of E into F. ∎

(3.1.6) Proposition

If a Fréchet space E admits a continuous norm and has the bounded approximation property then it is countably normed.

Proof

By (3.1.4), (3.1.5) it suffices to show that a space F which admits a continuous norm and has an FDD is countably normed.

Let $(A_n)$ be an FDD for F and $(||\cdot||_k)$ a sequence of norms defining the topology of F. As in (3.1.5) we define a sequence of norms $(|\cdot|_k)$ which also defines the topology of F by

$$|y|_k = \sup_n ||\sum_{i=1}^{n} A_i y||_k, \quad y \in F, \quad k \in \mathbb{N}.$$

Moreover, using standard ideas from elementary basis theory it is easy to see that $(A_n)$ is an FDD for each Banach space $(F,|\cdot|_k)^\wedge$.

Thus if $y \in (F,|\cdot|_{k+1})^\wedge$ and its image in $(F,|\cdot|_k)^\wedge$ is 0, then the series $\sum_n A_n y$ converges to 0 in $(F,|\cdot|_k)^\wedge$ so $A_n y = 0$, $n \in \mathbb{N}$. Then, of course, this series also converges to 0 in $(F,|\cdot|_{k+1})^\wedge$ so $y = 0$. ∎

(3.1.7) In view of (3.1.6), the problem of existence for BAP can be reduced to the problem of finding a space which admits a continuous norm but is not countably normed. This is difficult because we must find a space for which not every canonical map is 1-1, no matter how we choose the sequence of norms. To deal with this difficulty we must use an "invariant", that is a property which follows from countably normed and is independent of the choice of norms.

(3.1.8) A sequence of functions $A_k: S_{k+1} \to S_k$, $k \in \mathbb{N}$ is called weakly injective if

$$\forall x \in \bigcap_{k=1}^{\infty} A_1 \ldots A_k(S_{k+1}) \text{ there exists } (x_k) \ni x_1 = x \text{ and } x_k = A_k x_{k+1}, \quad k \in \mathbb{N}.$$

(3.1.9) The following diagram may motivate this definition:

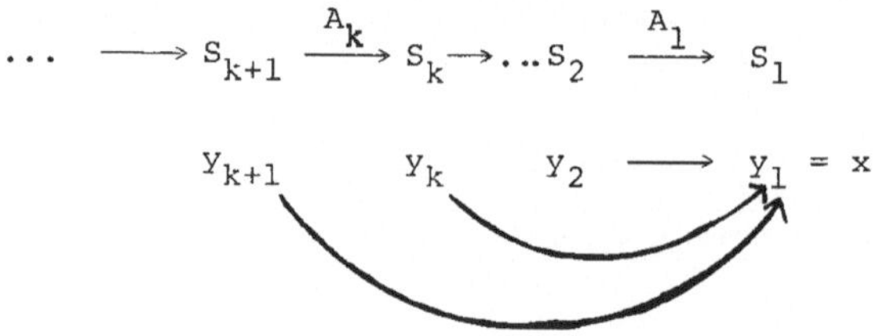

The hypothesis in the definition says that starting at x one can jump back k steps for any k, that is, there exists $y_k \ni x = A_1 \cdots A_{k-1} y_k$. The conclusion says that this can be done step-by-step, that is, $y_k = A_k y_{k+1}$. This is, of course immediate if each $A_k$ is 1-1 which explains the name we have chosen.

(3.1.10) Theorem

If E is a countably normed space, $(||\cdot||_k)$ a sequence of norms defining its topology with canonical maps $A_k$: $(E,||\cdot||_{k+1})^\wedge \to (E,||\cdot||_k)^\wedge$, $k \in \mathbb{N}$, then there exists $k_0 \ni (A_k)_{k \geq k_0}$ is weakly injective.

Proof

Let $(|\cdot|_k)$ be a sequence of norms defining the topology of E $\ni$ each canonical map $B_k$: $(E,|\cdot|_{k+1})^\wedge \to (E,|\cdot|_k)^\wedge$ is 1-1. Write $F_k = (E,|\cdot|_k)^\wedge$ and $E_k = (E,||\cdot||_k)^\wedge$.

Since $(||\cdot||_k)$ and $(|\cdot|_k)$ are two sequences of norms defining the same topology we have increasing sequences of indices $(j_k)$, $(\ell_k)$ and positive constants $(\alpha_k)$, $(\beta_k)$ such that

$$|x|_{\ell_k} \leq \alpha_k ||x||_{j_k} \leq \beta_k |x|_{\ell_{k+1}}, \quad x \in E,\ k \in \mathbb{N}.$$

We will show that $(A_k)_{k \geq j_1}$ is weakly injective.

Write $C_k = A_{j_k} \cdots A_{j_{k+1}-1}$, $D_k = B_{\ell_k} \cdots B_{\ell_{k+1}-1}$, $k \in \mathbb{N}$ so that each $D_k$ is 1-1. Now the above inequalities relate norms on E so the corresponding identity maps have continuous extensions. Thus we have the following commutative diagram:

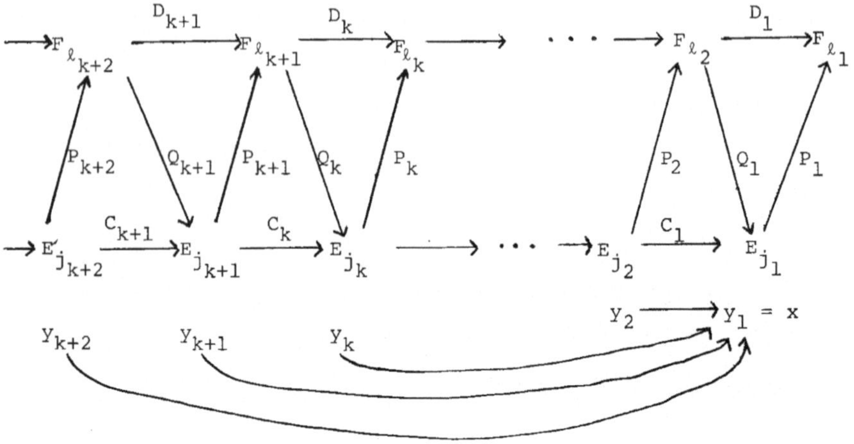


First we show that $(C_k)$ is weakly injective so we may assume $x = C_1 \cdots C_k\ y_{k+1}$ $k \in \mathbb{N}$. We define $x_k = C_k\ y_{k+1}$ so immediately $x_1 = C_1\ y_2 = x$. Also,

$$D_1 \cdots D_k P_{k+1} C_{k+1} y_{k+2} = P_1 C_1 \cdots C_k C_{k+1} y_{k+2} = P_1 x = P_1 C_1 \cdots C_k y_{k+1} =$$

$$D_1 \cdots D_k P_{k+1} y_{k+1}$$

and since each $D_k$ is 1-1, $P_{k+1} C_{k+1} y_{k+2} = P_{k+1}\ y_{k+1}$ sc $C_k\ x_{k+1} = x_k$. Hence $(C_k)$ is weakly injective.

Finally, if $x \in \bigcap_{k \geq j_1} A_{j_1} \cdots A_k(E_{k+1})$ then $x \in \bigcap_k C_1 \cdots C_k(E_{j_{k+1}})$ so we have $(y_k)$ with $y_1 = x$ and $y_k = C_k y_{k+1}$. Define $(x_k)$ by setting $x_{j_k} = y_k$, $k \in \mathbb{N}$ and $x_\ell = A_\ell \cdots A_{j_{k+1}-1} y_{k+1}$ for $j_k < \ell < j_{k+1}$. This shows that $(A_k)_{k \geq j_1}$ is weakly injective. ∎

(3.2) Now we construct a nuclear Fréchet space without the bounded approximation property by constructing a nuclear Fréchet space which admits a continuous norm but is not countably normed. Our construction is a projective limit of a sequence of operators $A_k$ on $\ell_2$. First we develop conditions on $A_k$ which guarantee all desired properties of the projective limit $\hat{E}$ and then we construct the infinite matrices determining the $A_k$ so that these conditions are satisfied.

(3.2.1) Let $A_k: \ell_2 \to \ell_2$, $k \in N$ be continuous linear operators and define

$$\hat{E} = \{(x_k): x_k = A_k x_{k+1},\ k \in N\},\ ||(x_j)||_k = ||x_k||.$$

Then it is easy to see that $\hat{E}$ with topology defined by $(||\cdot||_k)$ is a Fréchet space. It is, in fact, the projective limit of the maps $(A_k)$.

(3.2.2) Proposition

In the context of (3.2.1) assume that

i) $\ker A_k \cap \bigcap_{j>k} A_{k+1} \cdots A_j(\ell_2) = \{0\}$, $k \in \mathbb{N}$

ii) Each $A_k$ has dense range

iii) Each $A_k$ is trace class

iv) $\forall k \in \mathbb{N}$ there exists $x^{k+1} \in \ell_2$ such that

$$(x^{k+1} + \ker A_k) \cap A_{k+1} \cdots A_j(\ell_2) \neq \emptyset \quad \forall j > k$$

but

$$(x^{k+1} + \ker A_k) \cap \bigcap_{j>k} A_{k+1} \cdots A_j(\ell_2) = \emptyset \quad .$$

Then $\hat{E}$ is a nuclear Fréchet space with a continuous norm but it is not countably normed and hence $\hat{E}$ does not have BAP.

<u>Proof</u>

First we show that each $||\cdot||_k$ is a norm. Let $x = (x_k) \in \hat{E}$ and suppose $||x||_k = 0$. Then $||x_k|| = 0$ so $x_k = 0$. But $x_{k+1} = A_{k+1} \cdots A_j x_{j+1}$, $j > k$ and $A_k x_{k+1} = x_k = 0$ so by i) $x_{k+1} = 0$. Similarly $x_j = 0$ for all $j > k$ and clearly $x_j = 0$ for all $j < k$. Hence $x = 0$.

Now we define $\pi_k: (\hat{E},||\cdot||_k) \to \ell_2$ by $\pi_k x = x_k$. It is clear that $\pi_k$ is norm preserving and 1-1. We want to show that it has dense range. Let $\varepsilon > 0$ and $y \in \ell_2$. We define the quantities in the following layout:

$$
\begin{array}{ccccccccc}
\cdots & \longrightarrow & \ell_2 & \xrightarrow{A_{k+2}} & \ell_2 & \xrightarrow{A_{k+1}} & \ell_2 & \xrightarrow{A_k} & \ell_2 \\
 & & & & & & & & y \\
 & & & & & & x_1^2 & \longrightarrow & x_1^1 \\
 & & & & x_2^3 & \longrightarrow & x_2^2 & \longrightarrow & x_2^1 \\
\cdots & \longrightarrow & x_3^4 & \longrightarrow & x_3^3 & \longrightarrow & x_3^2 & \longrightarrow & x_3^1 \\
 & & \vdots & & \vdots & & \vdots & & \vdots
\end{array}
$$

That is, we define $x_n^j \in \ell_2$ for $j = 1,\ldots,n+1$ and $n \in \mathbb{N}$ in such a way that $x_n^j = A_{k+j-1} x_n^{j+1}$ and, because of ii) $||x_n^j - x_{n+1}^j|| \leq 2^{-n}\varepsilon$, and $||x_1^1 - y|| \leq \varepsilon$. It follows that $x^{k+j-1} = \lim_n x_n^j$ exists and $||x^k - y|| \leq 2\varepsilon$.

Hence if we define $x^\ell = A_\ell x^{\ell+1}$ for $\ell < k$ it follows that $x = (x^\ell) \in \hat{E}$ and $||\pi_k x - y|| \leq 2\varepsilon$.

Thus $\pi_k$ has dense range so it has a unique extension to an isometry $\hat{\pi}_k: (\hat{E},||\cdot||_k)\hat{\ } \to \ell_2$. It then follows that

$$(\hat{E},||\cdot||_{k+1})\hat{\ } \to (\hat{E},||\cdot||_k)\hat{\ } = \hat{\pi}_k^{-1} A_k \hat{\pi}_{k+1}.$$

In particular, by iii), $\hat{E}$ is then nuclear and the weak injectivity of the canonical maps would imply the weak injectivity of $(A_k)$. Hence by (3.1.10) we are finished if we show that for any $k$, $(A_j)_{j \geq k}$ is not weakly injective.

Given k we have $x^{k+1}$ from iv) so for $j > k$ we have $y_j$, $z_{j+1} \ni$

$$A_k x^{k+1} = A_k y_j, \quad y_j = A_{k+1} \cdots A_j z_{j+1} .$$

Hence $A_k x^{k+1} \in \bigcap_{j \geq k} A_k \cdots A_j(\ell_2)$ so if $(A_j)_{j \geq k}$ were weakly injective, we would have $(u_j)_{j \geq k}$ with $u_k = A_k x^{k+1}$ and $u_j = A_j u_{j+1}$, $j \geq k$. But then

$$A_k u_{k+1} = u_k = A_k x^{k+1}$$

so $u_{k+1} \in x^{k+1} + \ker A_k$ and, for $j > k$,

$$u_{k+1} = A_{k+1} \cdots A_j \; u_{j+1} \in \bigcap_{j>k} A_{k+1} \cdots A_j(\ell_2)$$

which contradicts the last statement in iv). ∎

(3.2.3) In this paragraph we construct the matrices which define the operators $A_k: \ell_2 \to \ell_2$, $k \in \mathbb{N}$. We begin by decomposing $\ell_2$ into infinitely many pairwise orthogonal infinite dimensional blocks, $H(\nu)$, $\nu \in \mathbb{N}$. Each $H(\nu)$ is similarly decomposed into subblocks, $H_m(\nu)$, $m \in \mathbb{N}$. For each $m, \nu \in N$ we fix an orthonormal basis $(e_{mn}(\nu))_n$. Thus, $\{e_{mn}(\nu): m,n,\nu \in \mathbb{N}\}$ is an orthonormal basis for $\ell_2$.

Let $\lambda_{mn}(\nu)$, $m,n,\nu \in \mathbb{N}$ be positive numbers such that

$$\sum_{m,n,\nu} \lambda_{mn}(\nu) < \infty .$$

Our first step is to define part of each $A_k$, $k \in \mathbb{N}$ as follows: If

$$\left.\begin{array}{l} \nu \neq k \\ \text{or} \\ \nu = k \text{ and } m < k \\ \text{or} \\ \nu = k = m \text{ and } n > 1 \end{array}\right\} \quad \text{let } A_k e_{mn}(\nu) = \lambda_{mn}(\nu) e_{mn}(\nu).$$

If

$$\nu = k < m \text{ and } n > 1 \qquad \text{let } A_k e_{mn}(\nu) = \lambda_{mn}(\nu) e_{m,n-1}(\nu).$$

Extending by continuity we then have that $A_k$ is defined on the subspace generated by the blocks $H(\nu)$, $\nu \neq k$; the subblocks $H_m(k)$, $m < k$; and a subspace of codimension 1 of each subblock $H_m(k)$, $m \geq k$. Moreover, on each block $H(\nu)$, $\nu \neq k$ the matrix of $A_k$ is diagonal with a non-vanishing diagonal sequence that converges to 0.

For convenience of notation we will write $A_{k+1} \cdots A_m$ for $m \geq k$ and consider this to be the identity operator when $m = k$.

Fix $m \geq k$. If we consider the operators $A_{k+1} \cdots A_m$, $A_{k+1} \cdots A_{m+1}$ restricted to the space $H$ generated by $e_{mn}(k)$, $n > 1$ then these two operators (as far as they have been defined) are invariant on the orthogonal complement of $H$ and on $H$ they are diagonal. Moreover if $(\alpha_n)$, $(\beta_n)$ are the respective diagonals then $0 < \alpha_n$, $0 < \beta_n$, $n > 1$ and $\lim_n \frac{\beta_n}{\alpha_n} = 0$. This implies that there exists $w_m^k \in H \ni w_m^k$ is in the range of $A_{k+1} \cdots A_m$ but not in the range of $A_{k+1} \cdots A_{m+1}$. We write

$$w_m^k = \sum_{n=2}^{\infty} s_{mn}^k e_{mn}(k).$$

Now $A_k w_m^k$ has been defined so we may write

$$s_{m1}^k = \frac{1 + ||A_k w_m^k||}{\lambda_{m1}(k)} \quad \text{and} \quad u_m^{k+1} = s_{m1}^k e_{m1}(k) + w_m^k$$

and define

$$A_k(e_{m1}(k)) = \frac{1}{s_{m1}^k}(e_{k1}(k) - A_k w_m^k) .$$

Notice that if this is done for all $k \in \mathbb{N}$, the properties of $A_{k+1} \cdots A_m$, $A_{k+1} \cdots A_{m+1}$ used above are not affected.

This completes the definition of the operators $A_k$, $k \in \mathbb{N}$.

(3.2.4) Proposition

In the context of (3.2.3) we have

i) $||A_k(e_{mn}(\nu))|| \leq \lambda_{mn}(\nu)$ for all $m,n,\nu \in \mathbb{N}$

ii) $A_k u_m^{k+1} = e_{k1}(k)$ for all $k \leq m$

iii) $u_m^{k+1} \in A_{k+1}\cdots A_m(\ell_2)$ but $u_m^{k+1} \notin A_{k+1}\cdots A_{m+1}(\ell_2)$ for all $k \leq m$.

Proof

i) This is immediate from the construction in all cases except $\nu = k \leq m$ and $n = 1$. For this case we compute

$$||A_k(e_{m1}(k))|| \leq \frac{1+||A_k w_m^k||}{s_{m1}^k} = \lambda_{m1}(k)$$

ii) From the construction we have

$$A_k u_m^{k+1} = s_{m1}^k A_k e_{m1}(k) + A_k w_m^k = e_{k1}(k) - A_k w_m^k + A_k w_m^k = e_{k1}(k)$$

iii) From the construction it follows that $e_{m1}(k)$, $w_m^k$ are both in $A_{k+1}\cdots A_m(\ell_2)$ so $u_m^{k+1}$ is also. On the other hand $e_{m1}(k)$ is in $A_{k+1}\cdots A_{m+1}(\ell_2)$ but $w_m^k$ is not so $u_m^{k+1}$ cannot be in. ∎

(3.2.5) Proposition

In the context of (3.2.3) we have

$$\ker A_k = \{v = \sum_{m=k}^{\infty} \xi_m u_m^{k+1} : \sum_{m=k}^{\infty} (\xi_m ||u_m^{k+1}||)^2 < \infty \text{ and } \sum_{m=1}^{\infty} \xi_m = 0\} .$$

Proof

If we have such a v then by (3.2.4) ii)

$$A_k v = \sum_{m=k}^{\infty} \xi_m A_k u_m^{k+1} = (\sum_{m=k}^{\infty} \xi_m) e_{k1}(k) = 0,$$

so $v \in \ker A_k$.

Conversely, if $v \in \ker A_k$ we first expand v as follows:

$$v = \sum_{\nu \neq k} \sum_{m,n} \eta_{mn}(\nu) e_{mn}(\nu) + \sum_{m=1}^{k-1} \sum_{n} \eta_{mn} e_{mn}(k)$$

$$+ \sum_{m=k}^{\infty} (\xi_m u_m^{k+1} + \sum_{n=2}^{\infty} \zeta_{mn} e_{mn}(k).$$

That is, on the blocks $H(\nu)$, $\nu \neq k$ and the subblocks $H_m(k)$, $m < k$ we have used the orthonormal basis, while on each subblock $H_m(k)$, $k \leq m$, since the component of $u_m^{k+1}$ on $e_{m1}(k)$ is $s_{m1}^{k} \neq 0$ we can replace $e_{m1}(k)$ by $u_m^{m+1}$ in the orthonormal basis $(e_{mn}(k))_n$ and we still have a basis.

Thus if we apply $A_k$ to v we obtain, from (3.2.4) ii)

$$0 = A_k(v) = \sum_{\nu \neq k} \sum_{m,n} \eta_{mn}(\nu) \lambda_{mn}(\nu) e_{mn}(\nu) + \sum_{m=1}^{k-1} \sum_{n} \eta_{mn} \lambda_{mn}(k) e_{mn}(k)$$

$$+ \xi_k e_{k1}(k) + \sum_{n=2}^{\infty} \zeta_{kn} \lambda_{kn}(k) e_{kn}(k)$$

$$+ \sum_{m=k+1}^{\infty} (\xi_m e_{k1}(k) + \sum_{n=2} \zeta_{mn} \lambda_{mn}(k) e_{m,n-1}(k)) .$$

But this is now an expansion in an orthonormal basis so the coefficient of each basis vector is 0. Hence we have, all $\eta, \zeta$ terms vanish so

$$v = \sum_{m=k}^{\infty} \xi_m u_m^{k+1}$$

which is an orthogonal expansion so $\sum_{m=k}^{\infty} ||\xi_m u_m^{k+1}||^2 < \infty$ and, taking the coefficient of $e_{k1}(k)$ we have $\sum_{m=k}^{\infty} \xi_m = 0$. ∎

(3.2.6) Theorem

The space $\hat{E}$ defined in (3.2.1) using the operators constructed in (3.2.3) is a nuclear Fréchet space which admits a continuous norm but is not countably normed and hence $\hat{E}$ does not have the bounded approximation property.

Proof

It suffices to verify the four conditions of (3.2.2).

Condition iii) is immediate from (3.2.4) i). For condition ii) it suffices to check that every $e_{mn}(\nu)$ is in the range of $A_k$. This is immediate from (3.2.3) if $\nu \neq k$, or $\nu = k > m$, or $\nu = k = m$ and $n > 1$ or $\nu = k < m$. Finally $e_{k1}(k)$ is the range of $A_k$ because of (3.2.4) ii).

Next we verify i). Let $v \in \ker A_k$ so by (3.2.5) $v = \sum_{m=k}^{\infty} \xi_m u_m^{k+1}$ and if $\xi_m = 0 \ \forall m > k$ then $v = 0$. But if we have $m > k$ with $\xi_m \neq 0$ then if $v \in A_{k+1} \cdots A_{m+1}(\ell_2)$ it would follow from the diagonality of this operator on each $H_m(k)$ and its invariance on the orthogonal complement of $H(k)$ that $u_m^{k+1}$ would be in the range of $A_{k+1} \cdots A_{m+1}$. But this is false by (3.2.4) iii). Thus i) holds.

Finally to verify iv) we set $x^{k+1} = u_k^{k+1}$, $k \in \mathbb{N}$. Let $v \in \ker A_k$ so by (3.2.5) $v = \sum_{m=k}^{\infty} \xi_m u_m^{k+1}$ and if $\xi_m = 0 \ \forall m > k$ then $x^{k+1} + v = (1+\xi_k) u_k^{k+1}$ which, by (3.2.4) iii) is not in the range of $A_{k+1}$. Thus if $x^{k+1} + v$ is in $\bigcap_{j>k} A_{k+1} \cdots A_j(\ell_2)$ we have $m > k$ with $\xi_m \neq 0$. But then, arguing exactly as in the previous paragraph we conclude that $x^{k+1} + v$ is not in the range of $A_{k+1} \cdots A_{m+1}$. On the other hand, for each $j > k$ we have $v = u_j^{k+1} - x^{k+1} = u_j^{k+1} - u_k^{k+1} \in \ker A_k$ by (3.2.4) ii) and $u_j^{k+1}$ is in the range of $A_{k+1} \cdots A_j$ by (3.2.4) iii) so

$$x^{k+1} + v = u_j^{k+1} \in (x^{k+1} + \ker A_k) \cap A_{k+1} \cdots A_j(\ell_2)$$

and iv) is verified.

■

## 4. Summary of Results

(4.1) We return now to the three problems listed in (1.3) and summarize the known results.

(4.1.1) The existence problem (1.3.1) is completely solved. Using the results in (1.2.1), (2.2.2) and (3.2.6) we see the answer is positive for the approximation property and negative for all other properties.

(4.1.2) For the permanence problem (1.3.2) we have, of course a positive solution for AP (1.2.1) and for FDD, SFDD, r-FDD, UPI, SUPI, r-UPI and basis the answer is negative in a very strong sense (2.2.1). The permanence problem for BAP remains open.

(4.1.3) Finally we consider the comparison problem. All implications in one direction are trivially true. Most (but not all) of the reverse implications are known to be false. These facts are contained in the statements and proofs of (2.2.3), (2.2.4) and (3.2.6). We display these results in the following diagram. A solid arrow represents an implication which holds and a dotted arrow represents one which fails. The facts which do not follow from our diagram by transitivity are unknown.

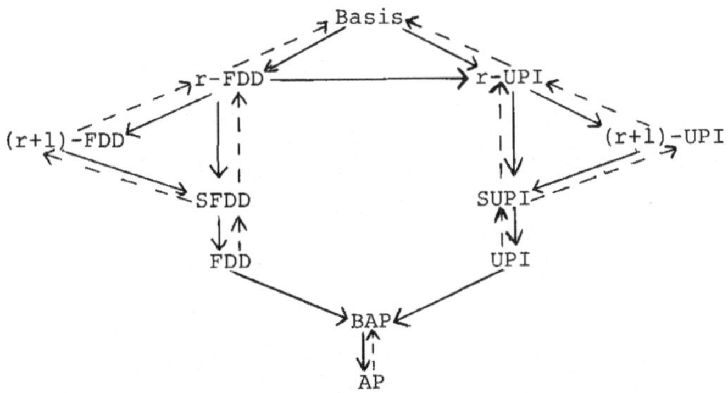


(4.1.4) The most interesting implications which remain open questions are: Does BAP imply FDD or UPI and does any of the UPI properties imply any of the FDD properties?

5. Notes and Remarks

This chapter contains almost all that is known about approximation properties in nuclear Fréchet spaces. The original ideas about bases, AP and BAP are due to Grothendieck [37]. The other approximation properties were originally considered for Banach spaces [38]. The analogues for nuclear Fréchet spaces (which are always separable) are quite natural.

There are important connections between the material in this chapter (especially section 2) and in Chapters III and IV. Indeed Lemma (2.1.4) which is a major tool in this chapter is really the same as II (2.2.3) and II (2.3.3) on which the main constructions of Chapters III and IV are based. In this sense one might say that some of the results in this chapter are an application of the preceeding material.

The statement in (1.2.1) is a standard result in the theory. A proof can be found in [57]. The statement in (1.2.3) is obvious.

The material in section 2 is a synthesis of ideas due to Mitiagin and Zobin [50], Djakov and Mitiagin [17], Mitiagin [48], Bessaga [5] the author [30], the author and Mitiagin [32], and the author and Bessaga [7]. The last paper is essentially the same as this section.

In Theorems (2.1.5) and (2.1.6) we have tried to make the two presentations as close as possible. Thus one can compare the two arguments and see exactly where the proofs for subspaces and quotient spaces differ. We feel that the differences are minor and in this sense we claim to have presented a unified proof for the two cases. It is also relatively short. For both points it might be interesting to compare with [30] and [32] which are the original proofs.

A little more is known about the counterexample in (2.1) which is the same as the space constructed by Djakov and Mitiagin [17]. Because of (2.2.1) and (1.2.3) this space is not a complemented subspace of any space with a UPI. Hence it is not a complemented subspace of a space with a basis. We do not know about the corresponding statements for FDD.

The main result on the bounded approximation property (Theorem (3.2.6)) is due to the author. A sketch of the proof appears in a research announcement submitted to the Proceedings of the U.S. National Academy of Sciences and the full details will appear in Studia Math.

The notion of countably normed space is due to Gelfand and Silov [36]. Proposition (3.1.5) is due to Pełczyński and Wojtarsczyk [52]. They actually prove that E is a complemented subspace of F. In view of the above remark it then follows that there exist nuclear Fréchet spaces E for which F cannot be nuclear. An even stronger version of (3.1.5) is true. Pełczyński has shown [51] that not only is E a complemented subspace of F, but F can be chosen to have a basis (see [44] for a detailed proof).

Proposition (3.1.6) was suggested to the author by A. Pełczyński in a private conversation.

The general idea of the construction of $\hat{E}$ is based on the author's notion of nuclear system studied in [23], [24], and [25]. It should be noted that there is a serious error in [24]. Indeed, Theorem 1 implies that every nuclear Fréchet space which admits a continuous norm is countably normed. This is, of course, false by (3.1.6). The error in the proof occurs on page 151, line 10 from the bottom where it is erroneously claimed (and subsequently used) that each $f_k$ is an isomorphism.

A little bit more is known about the permanence problem for BAP. For subspaces it follows from (3.1.4) that the counterexample constructed here is not a subspace of any space which admits a continuous norm and has a basis. Of course, by the result of Kōmura and Kōmure [41], our space is a subspace of the Cartesian product of countably many copies of (s) which has BAP. But the question of subspaces of a countably normed space with a basis remains open. It is not even known if every countably normed nuclear Fréchet space has BAP.

For quotient spaces it has just been shown that one of the spaces constructed in (3.2) is isomorphic to a quotient of (s). The details will appear in the Studia paper mentioned above.

## REFERENCES


1. M. Alpseymen, Basic sequences in some nuclear Köthe sequence spaces, Thesis, University of Michigan, 1978.

2. M. Alpseymen, N. De Grande-De Kimpe and Ed Dubinsky, Subspaces of $L_f(\alpha,r)$ spaces, in preparation.

3. A. Aytuna and T. Terzioglu, On certain subspaces of a nuclear power series space of finite type, Studia Math., to appear.

4. C. Bessaga, Some remarks on Dragilev's theorem, Studia Math. 31 (1968), 307-318.

5. C. Bessaga, A nuclear Fréchet space without basis. I. Variation on a theme of Djakov and Mitiagin, Bull. Acad. Polon. Sci. 24, 7 (1976), 471-473.

6. C. Bessaga, Nuclear spaces without bases: A proof of Dubinsky's theorem, Report to Functional Analysis Seminar, Univ. of Mich. (1978).

7. C. Bessaga and Ed Dubinsky, Nuclear Fréchet spaces without bases. III. Every nuclear Fréchet space not isomorphic to $\omega$ admits a subspace without a strong finite dimensional decomposition, to appear.

8. C. Bessaga and A. Pełczyński, An extension of the Krein-Milman-Rutman theorem concerning bases to the case of $B_0$ spaces, Bull. Acad. Polon. Sci. 5 (1957), 379-383.

9. C. Bessaga and A. Pełczyński, Własności Baz w Przestrzeniach Typu B, Prace. Mat. III (1959), 123-142.

10. C. Bessaga and A. Pełczyński, On embedding of nuclear spaces in the space of all infinitely differentiable functions on the line (Russian) Dokl. Akad. Nauk. SSSR 134 (1960), 745-748.

11. C. Bessaga, A. Pełczyński and S. Rolewicz, On diametral approximative dimension and linear homogeneity of F-spaces, Bull. Acad. Polon. Sci. 9, 9 (1961), 677-683.

12. L. Crone, Ed Dubinsky and W.B. Robinson, Regular bases in products of power series spaces, J. Funct. Anal. 24 (1977), 211-222.

13. L. Crone and W.B. Robinson, Every nuclear Fréchet space with a regular basis has the quasi-equivalence property, Studia Math. 52 (1975), 203-207.

14. L. Crone and W.B. Robinson, Diagonal maps and diameters in Köthe spaces, Israel J. Math. 20, 1 (1975), 13-22.

15. N. De Grande-De Kimpe, $L_f(\alpha,r)$-spaces between which all operators are compact, to appear.

16. N. De Grande-De Kimpe and W.B. Robinson, Compact maps and embeddings from an infinite type power series space to a finite type power series space, J. Für Math. Reine und Angew, 293/294 (1977), 51-61.

17. P.B. Djakov and B.S. Mitiagin, Modified construction of a nuclear Fréchet space without basis, J. Funct. Anal. 23 (4) (1976), 415-423.

18. M.M. Dragilev, Standard form of basis for the space of analytic functions (Russian), Uspehi Mat. Nauk 15 (1960) 2 (92), 181-188.

19. M.M. Dragilev, On regular bases in nuclear spaces, Amer. Math. Soc. Transl. (2) 93 (1970) 61-82.

20. M.M. Dragilev, Köthe spaces differing in diametral dimension (Russian) Sib. Mat. Z. 11, 3 (1970), 512-525.

21. M.M. Dragilev, Riesz classes and multiple regular bases (Russian) Functional analysis and theory of functions, Kharkov (1972), 65-77.

22. Ed Dubinsky, Perfect Fréchet spaces, Math. Ann. 174 (1967), 186-194.

23. Ed Dubinsky, Equivalent nuclear systems, Studia Math. 38 (1970), 373-379.

24. Ed Dubinsky, A new definition of nuclear systems with applications to bases in nuclear spaces, Studia Math. 41 (1972), 149-161.

25. Ed Dubinsky, Examples of nuclear systems, Studia Math. 42 (1972), 29-42.

26. Ed Dubinsky, Infinite type power series subspaces of finite type power series spaces, Israel J. Math. 15 (1973), 257-281.

27. Ed Dubinsky, Infinite type power series subspaces of infinite type power series spaces, Israel J. Math. 20 (1975), 359-368.

28. Ed Dubinsky, Concrete subspaces of nuclear Fréchet spaces, Studia Math. 52 (1975), 209-219.

29. Ed Dubinsky, Basic sequences in (s), Studia Math. 59 (1977), 283-293.

30. Ed Dubinsky, Subspaces without bases in nuclear Fréchet spaces, Journal of Functional Analysis 26, 2 (1977), 121-130.

31. Ed Dubinsky, Basic sequences in a stable finite type power series space, Studia Math., to appear.

32. Ed Dubinsky and Boris Mitiagin, Quotient spaces without bases in nuclear Fréchet spaces, Can. J. Math. 1978, to appear.

33. Ed Dubinsky and M.S. Ramanujan, On $\lambda$-nuclearity, Mem. Amer. Math. Soc. 128 (1972) 101 pp.

34. Ed Dubinsky and William Robinson, Quotient spaces of (s) with basis, Studia Math. 63 (1977), 39-53.

35. A.S. Dynin and B.S. Mitiagin, Criterion for nuclearity in terms of approximative dimension, Bull. Acad. Polon. Sci. III, 8 (1960), 535-540.

36. I.M. Gelfand and G.E. Silov, Quelques applications de la théorie des fonctions généralisées, J. Math. Pure et Appl. 35, 4 (1956), 383-412.

37. A. Grothendieck, Produits tensoriels topologiques et espaces nucleaires, Mem. Amer. Math. Soc. 16 (1955).

38. W.B. Johnson, H.P. Rosenthal and M. Zippin, On bases, finite dimensional decompositions and weaker structures in Banach spaces, Israel J. Math. 9 (1971), 488-506.

39. V.V. Kashirin, Subspaces of a finite center of an absolute Riesz scale which are isomorphic to an infinite center (Russian) Sib. Mat. Zhur. 16, 4 (1975), 863-865.

40. A. Kolmogorov, Über die beste annäherung von funktionen einer gebebenen funktionen, Ann. Math. 37, 1 (1936), 107-110.

41. T. Kōmura and Y. Kōmura, Über die einbettung der nuklearen Räume in $(s)^A$, Math. Ann. 162 (1966), 284-288.

42. G. Köthe, Topologische lineare Räume, Springer Verlag, Berlin-Gottingen-Heidelberg, 1960.

43. G. Köthe, Starke nukleare Folgenräume, J. Fac. Sci. Univ. Tokyo, sec. I 17 (1970), 291-296.

44. C. Matyszczyk, Approximation of analytic and continuous mappings by polynomials in Fréchet spaces, Studia Math. 60, 3 (1977), 223-238.

45. B.S. Mitiagin, Approximative dimension and bases in nuclear spaces (Russian) Usp. Mat. Nauk. 16 (4) (1961), 73-132.

46. B.S. Mitiagin, Equivalence of bases in Hilbert scales (Russian), Studia Math. 37 (1971), 111-137.

47. B.S. Mitiagin, Structure of subspaces of infinite Hilbert scales (Russian), Trudy 7 Simney Szkoly, Drogovic (1974), 127-133.

48. B.S. Mitiagin, Nuclear Fréchet spaces without bases. II. The case of strongly finite-dimensional decompositions, Bull. Acad. Polon. Sci. 24, 7 (1976), 475-480.

49. B.S. Mitiagin and G. Henkin, Linear problems of complex analysis, (Russian) Usp. Mat. Nauk. 26 (4) (1972), 93-152.

50. B.S. Mitiagin and N.M. Zobin, Contre-Exemple à l'existence d'une base dans un espace de Fréchet nucleaire, C.R. Acad. Sci. Paris, Ser. A. 279 (1974), 255-256, 325-327.

51. A. Pełczyński, Any separable Banach space with the bounded approximation property is a complemented subspace of a Banach space with a basis, Studia Math. 40 (1971), 239-242.

52. A. Pełczyński and P. Wojtarszczyk, Finite dimensional expansions of identity and the complementably universal basis of finite dimensional subspaces, Studia Math. 40 (1971), 91-108.

53. A. Pietsch, Nukleare lokalkonvexe Räume, Berlin, 1965.

54. M.S. Ramanujan and T. Terzioglu, Power series spaces $\Lambda_k(\alpha)$ of finite type and related nuclearities, Studia Math. 53 (1975), 1-13.

55. M.S. Ramanujan and T. Terzioglu, Subspaces of smooth sequence spaces, Studia Math., to appear.

56. S. Rolewicz, On spaces of holomorphic functions, Studia Math. 21 (1961), 135-160.

57. H.H. Schaefer, Topological Vector Spaces, Springer-Verlag, New York, Heidelberg and Berlin, Third Printing Corrected, 1971.

58. I. Singer, Bases in Banach Spaces I, Springer-Verlag, Berlin, 1970.

59. V.M. Tihomirov, On n-dimensional diameters of certain functional classes, Sov. Math. Dokl. 1 (1960), 94-97.

60. D. Vogt, Charakterisierung der Unterräume von (s), Math. Z. 155 (1977), 109-117.

61. D. Vogt, Charakterisierung der unterräume eines nuklearen stabilen potenzreihenraumes von endlichem typ, to appear.

62. D. Vogt and M.J. Wagner, Charakterisierung der Quotientenräume von (s) und eine Vermutung von Martineau, Studia Math., to appear.

63. D. Vogt and M.J. Wagner, Charakterisierung der Unterräume und Quotientenräume der nuklearen stabilen Potenzreihenräume von unendlichem typ, Studia Math. to appear.

64. M.J. Wagner, Unterräume und quotienten von Potenzreihenräumen, Dissertation, Wuppertal, 1977.

65. V.P. Zahariuta, On the isomorphism of Cartesian products of locally convex spaces, Studia Math. 46 (1973), 201-221.

66. V.P. Zahariuta, On isomorphism and quasi-equivalence of bases in Köthe spaces (Russian), Dokl. Akad. Nauk. 16 (1975), 411-414.

67. N.M. Zobin and B.S. Mitiagin, Examples of nuclear linear metric spaces without a basis, Funct. Anal. and Appl. 8, 4 (1975), 304-313.